T0344459

Seismic Resistance of Vernacular Timber Frames with Infills

This book provides an engineer's perspective on the traditional construction methods for timber frames with infills, focusing on traditional *paianta* houses in Romania and *minka* houses in Japan to provide insights into the construction, seismic behavior, and design considerations of such structures.

The nuances of each country's traditional construction methods are considered, as well as the shared seismic culture and the similar functionality and local materials used for the houses, plus challenges from earthquake loading and fire. The efforts to preserve traditional houses in Romania are contrasted with Japan's regulatory framework for traditional residential construction methods. Strengthening solutions are also proposed for timber-framed houses with infills, considering various causes of degradation.

- Introduces examples from non-seismic and seismic-prone countries.
- Provides a comparative analysis of worldwide examples.
- Presents design examples illustrating the integration of traditional architecture with modern design standards.

The book serves as a comprehensive guide to the engineering intricacies of traditional houses in Romania and Japan for engineers and architects, with practical applications for new constructions worldwide.

Seismic Resistance of Vernacular Timber Frames with Infills

Case Studies from Japan and Romania

Andreea Dutu and Yoshihiro Yamazaki

CRC Press is an imprint of the
Taylor & Francis Group, an informa business

First edition published 2024
by CRC Press
2385 NW Executive Center Drive, Suite 320, Boca Raton FL 33431

and by CRC Press
4 Park Square, Milton Park, Abingdon, Oxon, OX14 4RN

CRC Press is an imprint of Taylor & Francis Group, LLC

ISBN: 978-1-032-51704-9 (hbk)
ISBN: 978-1-032-52128-2 (pbk)
ISBN: 978-1-003-40537-5 (ebk)

DOI: 10.1201/9781003405375

Typeset in Times
by Newgen Publishing UK

Contents

Preface

This book delves into the exploration and documentation of traditional timber houses with infills in two distinct countries. Despite being separated by 9000 km, these houses share similarities in local materials and usage patterns. Visiting such homes evokes a common experience—one where the tangible connection to the past, history, and cultural heritage is palpable. Unfortunately, this shared heritage is often overlooked.

The focus of the book is on scrutinizing the nuances of each country's traditional construction methods, employing an engineer's perspective and considering the shared seismic culture. While both countries share seismic considerations, they differ in intensity and frequency. Examining traditional architecture worldwide reveals both similarities and inspirations from unique differences. Notably, smart construction details from one country can offer solutions for similar building types in another country, especially in post-disaster recovery scenarios.

The book briefly introduces examples from non-seismic and seismic-prone countries, emphasizing the specific characteristics of each type. It concludes with a comparative analysis of key details among worldwide examples, considering factors such as the system's name, number of stories, structural system per story, infill type, presence of braces, and connection types.

Traditional residential architecture in rural areas of Romania primarily relies on timber structures, featuring diverse layouts influenced by the geographical location (mountainous, fields, etc.), directly impacting the availability of raw materials. While adobe and cob houses exist in Romania, the focus is predominantly on timber-framed structures (*paianta*) due to their proven resilience over time, demonstrating durability in both seismic events and general wear.

Japanese traditional residential houses with timber frames, post and beam frameworks, and infills are presented. These structures have fascinated engineers for decades due to their refined construction details and simple structural layouts. The chapter highlights the transmission of carpentry skills through generations in Japan, a unique aspect not commonly observed worldwide. It explores the seismic behavior of traditional *minka* houses through real earthquakes and laboratory experiments.

Strengthening solutions are also proposed for timber-framed houses with infills, considering various causes of degradation. The importance of addressing the root cause before treating the effects is emphasized. The book sheds light on the efforts to preserve traditional houses in Romania, contrasting it with Japan's regulatory framework dedicated to traditional residential construction methods.

Ultimately, the book serves as a comprehensive guide to the engineering intricacies of traditional houses in Romania and Japan, providing insights into their construction, seismic behavior, and design considerations. Design examples are presented, offering practical applications for engineers and showcasing the integration of traditional architecture with modern design standards.

Acknowledgments

It took a long time to write this book, and I would like to thank my family for their patience and support. My husband, my mother, and my late father all encouraged me to go ahead and pursue my dreams no matter what. My two sons inspired me and empowered me. In fact, the book was started during my first maternity leave and was finished at the end of the second one.

I am thankful to all those from whom I learned in my career, and especially those who believed in me and never stopped doing so. I acknowledge all the opportunities my mentors offered me, and their kind collaboration despite the distance. It all started with Professor Joao Gomes Ferreira, who kindly welcomed me into his team and thus I started studying the topic of timber frames with various infills. He taught me how to write my first paper and showed me how a functional education system can weigh tremendously in a young researcher's life. But the most valuable thing he taught me was that supervisors can be friends with the students, even though their competence is way beyond the students'.

Professor Hiroyasu Sakata offered me the opportunity to go to Japan and be near greatness (in his lab), from which I learned and with which I lived, and thus my values and standards forever changed. It still is an honor to present myself as an ex-postdoc of Sakata laboratory. After my return to Romania, Professor Akira Wada never stopped teaching me greatness and many times lit my path that had become dark. A special thanks to my colleague and friend, Professor Mihai Niste, who helped and worked with me even though sometimes things got complicated. And also, to my international friends, Professor Shoichi Kishiki, Professor Fatih Sutcu, and Professor Zhe Qu, who keep me anchored in the real research world, which produces real and valuable results for society.

My co-author, Professor Yoshihiro Yamazaki, to whom I owe a significant part of my knowledge on both experimental and calculation methods, stood by me and helped from the first day I got to Japan and still, more than 10 years later, positively responds to my (sometimes) crazy research ideas.

Thanks to my Romanian friend, Eng. Valentin Nicula and Professor Dan Cretu, who kindly reviewed my design calculations in this book.

There are some others who can't be acknowledged but who deserve it to the fullest!

Without the help of Mr. Sometrey Mey and Mr. Ichiro Hirano with the experiments, I could not deliver the book on time. And I am also thankful to my friend, Arch. Daniel Mocanescu, the crazy inventor who constantly helps my ideas come true.

The research work behind this book, which allowed me to make a professional opinion on the topic, was supported financially by several research grants: project number PN-II-RU-TE-2014-4-2169" and PN-III-P2-2.1-PED-2021-1428 (tfmro.utcb.ro), both funded by UEFISCDI, Tokyo Institute of Technology's MSL funding and JSPS and CUEE postdoctoral fellowships.

About the Authors

Andreea Dutu is a structural engineer, who graduated with a bachelor's degree, master's degree, and PhD from the Technical University of Civil Engineering in Bucharest, Romania. Her main research background is related to the seismic evaluation of traditional timber buildings with various types of infills (masonry, clay, logs, etc.). She also has a passion for seismic isolation, energy dissipation, and vibration control of structures and she was elected as a member of the Board of Directors in the ASSISI Society (Anti-seismic Systems International Society).

Starting from her PhD studies, she has been enrolled in several research stages around the world (Portugal, France, Turkey, China), but the longest one was in Japan, at the Tokyo Institute of Technology, where she stayed for 2.5 years as a post-doctoral researcher. In Japan, under the supervision of Professor Hiroyasu Sakata and Professor Yoshihiro Yamazaki, her way of thinking profoundly changed from both a professional and personal point of view.

When she returned to Romania, she had a dream to apply what she had learned in Japan, meaning to never give up on pursuing her objectives (to do research as in Japan) and to embrace trial and error and accept it as a way of learning and growing. So far, she has obtained and successfully implemented three research grants in her home university where she also teaches several courses and is able to study traditional houses from the engineering point of view for the first time in Romania, conducting an extensive experimental program on sub-assemblies (walls, connections, component materials, etc.).

Yoshihiro Yamazaki is a structural engineer, who graduated with a bachelor's degree from Tokyo Metropolitan University and a master's degree and PhD from the Tokyo Institute of Technology in Japan. After earning a PhD he became an assistant professor at Tokyo Tech.

In his doctoral dissertation, he conducted research on design methods to reduce torsional vibration by applying passive dampers to wooden houses. Since then, he has continued his research on the seismic performance of wooden buildings. Recently, he has been conducting research and development on the realization of mid-to-high-rise wooden buildings to promote the use of wood in the construction industry in response to the growing awareness of environmental issues. Professor Yamazaki began researching timber-framed masonry structures with Andreea while she was in Japan. He has also been working on hybrid structures that combine concrete and steel members with wood.

At his current job at the Building Research Institute, he is working on the revision of Japanese technical standards for building structures, especially for CLT panel construction.

Introduction 1

When we visit a foreign country, we all like to see the local culture. What better way to observe it than through its traditional houses? In these houses, well preserved especially in the village museums, you can notice the habits of local people, the way they organize their lives through spaces inside the houses, and the aesthetic sense, which is very different from country to country.

Nowadays, most people wish for new materials for their new houses, and this is a sound reason to actually use the recent research results on new technology and materials. Other people want to learn from tradition and use it in a modern and more practical way in the present, taking advantage of the new technologies applied to natural materials, such as insulation panels, rammed earth, cob, etc. But in order to learn from tradition it is necessary to understand it first.

In this book, I chose to focus on and describe two completely different countries' traditional timber houses. The local materials used for these houses are not very different, although 9000 km apart. The way people use them is also not very different, and the feeling one gets while visiting such a house is, in fact, the same: you can actually touch the past, the history, the baggage that made us who we are nowadays. However, we have a tendency to forget it. This book is about noticing the details of each traditional construction method and comparing them, with an engineer's eye and taking into account the local seismic culture, which both countries have in common, although not at the same intensity and frequency.

Romania is a moderately seismic country, but one of the most affected countries in Europe by this hazard type. The last big earthquake occurred in 1977 and had a magnitude of 7.2 on the Richter scale. The characteristics of that event showed that the Vrancea seismic source usually produces earthquakes with long periods which affect the flexible or tall structures (many high-rise reinforced concrete buildings in Bucharest collapsed).

Japan is a highly seismic country, famous for its engineered buildings which could withstand the Oki Tohoku Earthquake in 2011, with a magnitude of 9.0 on the Richter scale, with no significant damage. Earthquakes are very frequent in Japan, and I personally can state that during my stay there,

DOI:10.1201/9781003405375-1

between May 2012 and September 2014, there were many months with weekly earthquakes, most likely aftershocks of the 2011 earthquake.

Then, looking at the world's traditional architecture, we find many similarities but also get inspired by the particular differences, and in some situations (i.e. recovery after disasters) the smart construction details in a country may save buildings with similar typologies in another country.

Traditional Architecture Worldwide

2

2.1 INTRODUCTION

Traditional timber frames with various types of infills can be found all over the world, and although they sometimes are poorly executed, in many situations they surprised engineers with their resilient behavior in earthquakes. They do not go through earthquakes without any damage, but usually, they do not collapse, and they protect the life of their inhabitants.

Maybe the oldest example is the "*Opus Craticium*" [1], which can be still found in the archeological site in Herculaneum. It was designed by Vitruvius, and it was a construction method used by the Romans, probably adapted from other cultures. Vitruvius also wrote one of the few texts on such type of construction and explained the wall building methods, but also the disadvantages one may experience during construction, such as detachment between the timber frame and the masonry infill, fire considerations, and moisture influence [2].

In Romanian territories, it seems the appearance of timber frames with infills ("*paianta*") dates back to the end of the early Neolithic [3]. Similar structures were identified in archeological sites dating from the Bronze Age and the Middle Age. Other countries, such as Italy and Greece, have similar records [4, 5].

The structural components of this typology are the timber skeleton infilled with various types of materials, which can be masonry, earth mixed with straws, wattle and daub in different layouts, and timber logs. In some countries, the houses also have timber diagonals, which usually increase the lateral loads' strength capacity of the building up to two times [6]. The timber has a confining effect for the infills, increasing the shear capacity of the infills, which otherwise may be subjected to a brittle failure (in the case of masonry).

DOI:10.1201/9781003405375-2

At the same time, the infills bring an increase in stiffness to the timber skeleton, which otherwise has a significant deformation capacity. These principles are the same for most of the houses of this type, no matter the country they are found in, but when we go into details, there are many variations, depending on the local availability of materials, skills of the workmanship, and also seismic culture. Over time, and with earthquakes "testing" them, people have been able to see which details are better for seismic resistance or not.

It is remarkable to notice that almost all the countries in the world have some traditional house similar to this system. Probably due to the ease of building and also the aesthetics of this architecture, such timber frames with infills can be found in countries with no seismic activities, such as Germany (*fachwerk*), France (*colombage*) (Figure 2.1), Scandinavia (*bindingverk*), and the United Kingdom (*half-timber*) [2].

Timber frames with infills can also be found in seismic countries, and most of them have shown resilient behavior, so even if they suffer some damages, they are easy to repair and repairs can be done with the same materials. When the infill collapses, for instance, it can be reused in the same way. Thus, such traditional houses can be found in Turkey (*himis*) (Figure 2.2a), Romania (*paianta*) (Figure 2.2c), Bulgaria (*Паянта*) (Figure 2.2d) and India (*dhajji-dewari*) [7]. Places like Portugal (*Pombalinos* buildings), Italy (*Casa Baraccata*), and Greece (*Xylopikti Toichopoiia*), as well as in America such as Salvador (*Bahareque*), Paraguay (*estaqueo*), Haiti (*gingerbread* and *kayi pey*) and Perú (*Quincha*) [8].

Some of the timber-framed masonry (TFM) houses, as they were called in [9], due to the predominant masonry infill, are surprisingly based on engineering knowledge and were enforced by law by the governments of those times. This is the case for Portugal, Italy, Greece and Peru.

FIGURE 2.1 Timber frames *colombage* in France Strasbourg.

FIGURE 2.2 Timber frames in Europe; some cases use plaster of mud or lime; *Entramados* Spain (a); *Himis*, Turkey (b); *paianta*, Romania (c); Паянта, Bulgaria (d).

Further in the book for each country known to possess such a type, the TFM house is briefly described. Thus, in the end, we can have an image of what each country has specifically and what is common for all of them. It is important to notice the particularities, because after a major earthquake, if reconstruction is needed, the specific details make all the difference, as long as we point them out and remember them.

2.2 TYPES OF TIMBER FRAMES WITH INFILLS AROUND THE WORLD

2.2.1 Portugal – *Pombalino* Buildings

After the 1755 earthquake and tsunami that affected the center of Lisbon, the Marquis of Pombal, the prime minister at that time, ordered the reconstruction with a solution to use the large amounts of debris left from the damaged unreinforced masonry houses. After the investigation of the affected houses, the engineers and architects came up with plans that included references to the height and orientation of the building and structural system, and thus appeared one of the first seismic design codes in Europe.

That's why the building was called "*Pombalino*", and it is still nowadays widely spread on Portuguese territory, especially in Lisbon. The structural system consists of a timber frame, which includes St. Andrew's crosses.

The "*Pombalino*", one of the most emblematic structures of its time, was the result of incorporating traditional timber frame structures with diagonal braces into its construction design to enhance its seismic capacity. This design is characterized by an exterior masonry wall, and it incorporates timber elements in such a way as forming a "cage". As a result of this innovative approach, "*Pombalino*" can achieve heights of up to five floors [10].

Emerging in Lisbon as part of the post-1755 earthquake city rehabilitation efforts, this structural typology primarily aimed to address the contemporary housing crisis. In this scenario, the use of timber for earthquake-resistant construction was linked to the knowledge gained from shipbuilding, as timber demonstrated exceptional resistance to horizontal forces. Concurrently, masonry, being fire-resistant, complemented this approach, resulting in a structural system characterized by timber frames confining masonry infills. An interesting and ingenious way to clear the city from the remaining debris after the earthquake was using it as a masonry infill for the newly reconstructed buildings.

Its foundations consist of timber piles of varying depths depending on the soil type in the construction area. Atop these piles, the structure supports a ground-level stone masonry floor, walls, and pillars interconnected via an arch system. Above this foundation level, a three-dimensional grid is created, showcasing an innovative approach featuring composite lattice walls known as "frontals", renowned for their load-bearing capabilities and on which several experimental programs were developed, one of them in which I participated under the supervision of Professor Joao Gomes Ferreira [11–15]. These structural interior walls consist of timber frames with both vertical and horizontal elements, reinforced by diagonal components (resembling St. Andrew's crosses), and filled with masonry to enhance their resistance against compression, tension, and bending forces. Traditional joints and iron nails connect these timber elements to the flooring structure, forming a three-dimensional timber framework characterized by increased stiffness and flexibility when subjected to seismic forces [16].

In the 18th century, the "frontal walls", represented an advanced antiseismic construction system, a structural system made of walls arranged in an orthogonal mesh of vertical panels (frontal walls) connected to timber floors. Each panel was further subdivided into triangular shapes, through its timber skeleton. The interconnection of these components played a vital role in strengthening the structure and transmitting vertical loads to the ground floor. Some studies suggest that these walls were effective in reducing the risk of out-of-plane collapse [17].

The floors were constructed with primary and secondary beams, arranged orthogonally in a regular geometric pattern. These beams were supported by vertical structural elements, including frontal walls or partition walls, which not only carried vertical loads but also absorbed horizontal seismic forces [18].

Furthermore, the ceilings adhered to a traditional design, featuring trusses, edgings, boards, and beams symmetrically arranged. These elements were directly supported either by stone masonry or external timber walls [19].

In Pombal's urban plan, buildings of this type were strategically grouped into rectangular and uniform blocks (measuring 702 × 5 m^2), each with a central garden. This layout aimed to ensure the consolidation of these houses as a cohesive unit in the event of a seismic event, promoting consistent and uniform behavior among them. These typical four-story buildings had mixed-use, with commercial spaces on the ground floor and residential levels above.

2.2.2 Italy – *Casa Baraccata*

Following the earthquake of 1783 in Sicily, Italy, which devastated a significant portion of the city, the Bourbon government drew upon adaptations and studies of the region's traditional construction methods [20]. They subsequently introduced a seismic code known as "*Istruzioni per gli Ingegneri Commissionati Nella Calabria Ulteriore*", which provided engineering guidelines for building design.

One prevalent type of structure, the *Baraccata* house, featured an internal timber frame with concealed diagonal braces, not visible from the outside, and covered by external masonry walls. This dual system connected the timber frame and unreinforced masonry exterior walls through metal joints, with special emphasis on corner connections to confine the masonry. Additionally, the floors and ceilings were constructed of wood, enhancing the structural stability and preventing overturning mechanisms. This structural system was engineered to withstand seismic and static forces in all directions, with nodes reinforced using slits and wooden pieces (or metal nails) to ensure high ductility. During the 1908 earthquake in Calabria, this building type demonstrated its safety and effectiveness as an anti-seismic solution [21].

Originating in Calabria, Italy, these traditional systems emerged after the 1783 earthquake and were incorporated into building codes as seismic solutions. The main structure consisted of frames affixed to a sturdy platform or connected to a robust base (mudsill) made of oak. The primary support posts ran continuously from the foundation to the ceiling, while smaller studs were elevated 0.80 cm above the ground in which horizontal elements form the floor platform, to avoid moisture problems, and some treatments were carried out on the main posts before being buried into the foundation. The timber frame included posts, ties, and braces in specific locations, sometimes forming structures like St. Andrew's crosses. The walls had an approximate thickness of 30 cm, which included the final plaster covering for both the exterior and interior walls. Frames with windows and doors had additional confining structures [22].

Horizontal elements, such as floors and ceilings, featured beam heads extending outward to rest on the beams of the lower floor frames. Attention was paid to the roof structure and foundations, as they were susceptible to water infiltration, a primary risk for timber deterioration.

Notably, Bianco's study highlighted the *Baraccata* house's structural configuration, emphasizing unification as a single structure through the inclusion of diagonal bracing throughout the construction. These braces provided lateral (in-plane) resistance, and the regularity in dimensions, both in plan and elevation, enhanced its earthquake resilience.

Primarily intended for housing, these structures were typically townhouses, often with biaxial symmetry and one or two floors. The symmetrical arrangement of frames reduced the risk of collapse. Following Bourbon's seismic resistance guidelines, the size of the facade was also restricted to ensure stability.

2.2.3 Greece – *Ξυλοπηκτη Τοιχοποιια (Xylopikti Toichopoiia)*

The earthquake's impact on the Island of Lefkada in Greece in 1821 prompted the British authorities, who were governing the Ionian Island at the time, to establish specialized construction codes for seismic-prone areas. In 1827, they officially published a building code that incorporated the key characteristics of the traditional island constructions. This code provided guidelines on construction materials, masonry wall thickness, maximum building height, and the spacing between structures [23]. As a result, a structural system was developed that primarily relied on a box-like configuration between its walls and floor diaphragms. Additionally, it included a secondary support system comprising timber columns positioned near the ground floor's masonry walls to carry the loads from the upper floors in cases of damage or collapse during earthquakes. This ensured the efficient transfer of vertical loads. Furthermore, half-timbered walls incorporated diagonals and crossbeams to enhance their confinement. On the upper floors, they were connected to the ground floor's masonry using variously sized and shaped metal elements by means of timber beams. The design of this structural system was primarily based on its ability to withstand displacement rather than just resisting forces [24].

Following the earthquake in 2003, a study conducted by [25] revealed that houses built with this structural system, even when poorly maintained, exhibited excellent performance during multiple seismic events.

2.2.4 Peru – Adobe-*Quincha* Structures

After the 1609 earthquake, one of the first references was issued in Peru, which regulated for the first time the use of *quincha* houses for the upper stories, vaults and domes [2]. Then, the 1746 earthquake stroke, producing many casualties and damages, so the "Conde de Superunda" assigned a mathematician, Luis Gaudin, to study the earthquake behavior of houses, and issued "The rules of good construction", which include the rules established in the previous reference and also established the use of high timber walls coat with board and muddy cane covers or fibers called *quincha*, to increase the seismic resistance, as a mandatory requirement for new construction [26].

Quincha houses prevail in coastal areas, and their ground story is made of adobe masonry walls resting on stone foundations, while the upper stories have timber skeletons and as infill a latticed timber system, which is actually the *quincha*. This infill is an interesting mixture of bricks and wattle and daub, which are covered with an earth layer and plaster at the end.

When the Spanish arrived, due to the often earthquakes, the original *quincha* technique was adapted, thus braces were added as a countermeasure, and also brick masonry infill in the bottom part of the walls.

The structural system of a *quincha* wall is composed of timber posts, horizontal elements (beams), diagonal elements and infills. The usual distance between the posts is between 0.45 m to 1 m, and the connections to the beams are mortise-tenon type with dowel. The walls present also small brace elements, at the bottom part of the walls, having a height of 0.7 to 0.9 m, and they are usually nailed between the posts and the bottom beams. Up until these braces' level, the infill is made of bricks or adobe, while the rest of the wall is made of wattle and daub.

2.2.5 Turkey – The *Himis* and *Bagdadí* Structures

In the aftermath of the 1509 earthquake in Istanbul, Turkey, the Ottoman government took measures to prohibit stone masonry construction and instead encouraged the development of timber-frame dwellings. By the end of the century, this initiative had transformed the city into a landscape dominated by timber-frame constructions [7]. This traditional architectural style was not driven by prescribed regulations but rather by the seismic wisdom of its builders.

In Turkey, this traditional Ottoman timber frame, known as *Himis*, consists of a foundation and a ground floor made of rubble stone masonry. Regularly spaced timber posts support the upper framework, which can be filled with non-load-bearing masonry or left open. While the ground floor often accommodates uneven terrain, the upper floors exhibit a more regular layout, emphasizing the "sofa room" as a central space crucial to the house's spatial distribution. The upper floors feature braced frames with integral vertical, horizontal, and diagonal components, all in one piece. The timber elements' cross-section typically measures 12 to 15 cm, with intervals between posts ranging from 1.30 to 1.50 m. Diagonal braces are inclined between 30° to 45° and are typically positioned to reinforce the marginal/corner posts of the frame [27].

Another architectural style, *Baghdadi*, is characterized by a reinforced timber frame with surfaces joined by a network of slats, both on the interior and exterior faces. These slats provide support for potential plaster or earth mortar. The interior of the frame can be infilled with lightweight materials such as timber pieces. The use of timber laths enhances resistance to lateral forces, making *Baghdadi* construction highly earthquake-resistant. However, it is more susceptible to natural deterioration and requires a significant amount of timber compared to more readily available materials like stone and earth. Consequently, it was less commonly used. Both *Himis* and *Baghdadi* structures gained prominence between the 15th and 19th centuries in response to numerous earthquakes in Turkey. Although they were later banned by the government due to fire-related safety concerns, these traditional construction methods have seen a recent resurgence owing to their exceptional seismic performance. Furthermore, these structures have been the focus of experimental and numerical studies, particularly regarding their frames [28].

The selection of timber species for such structures is contingent on the local supply and available resources. The choices encompass pine, oak, chestnut, hornbeam, juniper, cedar, fir, and poplar. Pine stands out as the most commonly employed timber, while white oak, chestnut, and cedar were reserved for relatively upscale residences. Poplar, on the other hand, was primarily utilized for roofing. The connections between these timber elements were typically fashioned using simple nail details for cross or section connections and mortise-and-tenon joints for attaching the main framework elements to the mudsill or the projecting components on upper floors.

2.2.6 Pakistan – *Dhajji-Dewari* Structures

Originating in India, Pakistan, and the Himalayas, the *Dhajji-Dewari* structure represents a reinforced timber frame filled with earth and stone bricks. The term "*Dhajji-Dewari*" originates from Persian, signifying a "quilted mosaic wall." Within this patterned framework, a network of timber bars is situated at various angles between the posts, creating a mosaic of masonry within the frame.

Typically, *Dhajji* buildings possess a multi-story rectangular design, often resulting from various alterations over time. The distribution of window openings and doors on the facade does not exceed 30 percent of the total area. Variations in construction are influenced by flat or sloping terrain, potentially introducing higher torsional responses and asymmetric plans under seismic excitation. The primary vertical support system is the timber frame, with the back-placed padding not designed for vertical loads. Additionally, wood contracts over time under permanent loads, contributing to the structure's stability during out-of-plane events [19].

The structural arrangement includes the separation of masonry on each upper level through a straight timber staircase-like interface. This design allows energy dissipation in case of masonry movement due to horizontal forces, minimizing cracking and increasing stiffness with each added floor, depending on the membrane separating each story [29].

Lateral load resistance in this system results from the combination of the timber frame and masonry infill (stone or brick) using weak mud mortar. This mortar quickly cracks under lateral loads, absorbing energy through friction between the infill and the timber frame. The main timber frame and braces behave elastically, preventing cracks from propagating through the infill. This configuration enables a reduction in wall thickness, lowering the building's mass and, consequently, the inertial forces it must withstand. Within this system, columns connect to primary beams, which are further linked to secondary beams forming the floor, with connections secured by nails. The intermediate floor, consisting of beams and an earthen floor, distributes lateral loads to the wall system but is not entirely rigid.

2.2.7 China – The *ChuanDou* Buildings

The *ChuanDou* vernacular buildings are spread in China, and some of them are infilled with masonry while others are not. They have a timber frame composed of posts which rest directly on the ground or on a stone, and tie beams connecting the posts through mortise-and-tenon joints. No braces are included in the timber frame system.

In [30], Professor Qu investigated the behavior of such houses after the Lushan earthquake in 2013 and found that such traditional houses exhibit damages, but rarely collapse. In general, they behave better than URM houses, but less than reinforced masonry ones. The field study also showed that even if masonry infills collapse, the lateral load can be carried by the timber frame alone [31].

This result inspired further research of such houses, and experimental tests were conducted by [32] on *ChuandDou* with/without wattle and daub infill, by [33] on timber frames with/without wallboard infill and by [34] on timber frames with and without brick masonry infill.

2.2.8 Nepal – The *Centibera* Buildings

The *Centibera* buildings feature stone masonry walls up to the plinth level or the first floor. Above this, the walls are constructed using wooden frames (posts and beams), infilled with woven bamboo mesh, and plaster made of cement mortar or a mixture of mud, cow dung, and rice husk [35]. Following the earthquake in September 2011, it was observed that *Centibera* buildings remained crack-free. In contrast, nearby brick masonry buildings displayed numerous cracks in their masonry walls. An on-site investigation of buildings in the Kathmandu valley revealed that the predominant architectural structural system in Nepal involves a timber-masonry composite structure. The timber column and beam elements are often concealed within the masonry walls, making them not visible from the exterior [35].

2.2.9 Haiti – The *Gingerbread* and *Kayi Pey* Houses

The Haitian *Gingerbread* houses can have three primary construction systems: braced timber frame, *colombage* (braced timber frame with masonry infill, as the French ones), and masonry bearing wall. Often, these construction methods are combined, resulting in hybrid typologies [36].

Braced timber frame construction consists of vertical posts that are connected through mortise-and-tenon joints to the mudsills and top plates of each story, connected using timber dowels. Diagonal timbers are strategically placed at corners and other locations to provide bracing for the frame assembly [36].

The timber frame with masonry infill, known as "*kayi pey*", are the vernacular type of Haitian house and features construction details that are rudimentary and lack engineering influence.

The *colombage* building technique, as its name suggests, closely resembles the French approach, with a timber frame filled with masonry. This timber framing incorporates top plates, mudsills, vertical posts, and diagonal braces, often employing mortise-and-tenon joinery in the frame's construction [36].

2.2.10 Japan

Japan doesn't usually have timber frames with masonry infills, but only a few cases of Western-influenced architecture, especially in mills, an example being the Tomioka Silk Mill in Gumna prefecture. For this particular building, experimental tests were conducted on a shaking table and [37] studied strengthening solutions for the existing building. An important finding was that the uppermost layer of mortar, between the infill and timber beam is very important to reduce the out-of-plane deformation of the walls. AFRP rods were then introduced in the mortar joints of the masonry, and this solution proved to increase the earthquake performance of the building.

Other studies were done on the existing buildings [38, 39], many of them considered heritage, having this structural system.

2.2.11 Ecuador – the *Bahareque* Buildings

The *Bahareque* construction method involves a combination of various materials, including *guadua* bamboo, timber, mud, straw, cow manure, and plaster. While there are numerous variations, the most common form in Ecuador features a framework constructed from bamboo or sturdy timber (such as mangrove, oak, or lignum vitae, chosen for their strength and durability). This framework includes posts anchored in a compacted soil or concrete foundation, with bamboo or wooden laths attached to it. The infills are made with mud, often mixed with straw, to create a continuous wall surface [40]. Low-rise (one- or two-story) *Bahareque* buildings that are well-connected and have been properly maintained to prevent deterioration of the plant-based materials have demonstrated good performance under seismic conditions [41].

Another variant of traditional, non-engineered construction in Ecuador uses clay brick infill instead of plastered laths in timber frames given the greater durability of clay masonry. In this construction system, it is common to include diagonal timber members in addition to the vertical and horizontal posts for greater strength and stability.

2.2.12 Myanmar

Timber frames are widely observed in Bagan and Chauk [42]. And after the earthquake in August 2016, such houses were considered to have performed well, because the observed damages were only as out-of-plane falling of the infills in some cases. However, no full collapse was observed for such a house [43].

2.2.13 Bulgaria – *Паянта*

Bulgaria, although it does not have a high seismicity, sometimes feels the Romanian earthquakes. And due to the common history with Romania and other Balkan countries, the vernacular houses with timber frames and infills share similar characteristics, and some of them have seismic timber bands in the stone masonry, while others look exactly like the Romanian *paianta* houses shown in Figure 2.2d.

2.3 MATERIALS AND GEOMETRIES' COMPARISON

The layout of the timber frame is given by the position of the windows and doors, thus creating the openings that impose the location of two posts and the top plate. The corner posts give the limits of the house, and in between the openings' posts and corner ones, intermediary posts are placed at a regular distance of more or less 1 m. When the posts are placed, the necessity of diagonal (bracing) elements arises, to support the posts into a vertical position. Some horizontal stiffening elements can also be present, when necessary. The frame thus created can be infilled with whatever the owner finds easier, cheaper or local. It may be stone, brick, or adobe masonry (depending on each territory). The infill influences greatly the seismic behavior of the house, in terms of strength and stiffness, and lime mortar and fired clay bricks make a stiff infill, while the mud/earthen mortar is usually cracking even during setting, and thus, it is usually not perfectly attached to the frame. For the later type, infill's influence is reduced, but still much better than without infill.

The *type of connection* between timber elements depends on the available material, the skills of the workmanship and also the cultural identity of the area. In TFM structures the joints are as important as the layout of the timber elements. The *carpentry joints* are usually suitably shaped to transmit forces by direct contact and friction. Their configurations and joints are usually complex and prove the level of craftsmanship and a good understanding of the structural behavior that has resulted from the evolution of construction systems [2]. The collaboration between the infill and the frame is the secret of the resilient behavior of such houses in seismic events, the infill panels offer rigidity, working as a rigid body with the ductile frame, which deforms, but not enough to produce the collapse of the entire house.

The most common traditional timber joints in the timber-framed houses with infills are the mortise-and-tenon ones (Figure 2.3a), which may have a nail or two in the center of the connection, or even a timber dowel; and then, the cross-halved ones (Figure 2.3b), often called half-lap joints, which may also have a nail or two in the center. In the old times the nails used to be made of iron, but newer examples have steel nails, and the recent ones even screw nails. Other types of joints can also be found, but not as often as the first two, such as the stepped joints (Figure 2.3c), dovetail (Figure 2.3d), and scarf joints useful to connect beams (mudsills or top plates) (Figure 2.3e).

In Table 2.1 a comparison was made between the worldwide traditional timber frames, in terms of different details, such as the number of stories, infill types, if they are equipped with braces and also connection types.

FIGURE 2.3 Traditional timber joint types: mortise-and-tenon, with or without timber dowel or nails (a); cross-halved with nails (b); stepped (c); dovetail (d); and scarf joint (e).

TABLE 2.1 Comparison of the main construction details of the infilled timber frames worldwide

SYSTEM	*LOCATION*	*STRUCTURAL MATERIAL*		*NO. OF FLOORS*	*INFILL TYPE*	*BRACES*	*CONNECTION TYPES*
		GROUND FLOOR	*UPPER STORY*				
Pombalino	Lisbon, Portugal	Masonry bricks	Timber frames+ masonry	3 or more	Stone / brick	O	Cross-halved, nails, mortise-and-tenon
Baraccata	Bourbon, Italy	Masonry bricks/ timber	Timber frames+ masonry	2	Stone / brick	X	Cross-halved and nails
Dolomites	Northern, Italy	Masonry	Timber frames	2	Branches+plaster/ Stones+plaster/ Bricks+plaster/ Stone	X/O	Dovetail
Half-Timber	Canterbury, England	Masonry bricks/ timber	Timber frames decorative exterior+ masonry	2 or 3	Stone / brick (plaster exterior)	O	Cross-halved
Colombage	France	Masonry	Timber frames (decorative) exterior+ masonry	2 or 3	Stone / brick (plaster exterior)	Δ	Mortise-and-tenon
Ξυλοπηκτη Τοιχοποιια (Xylopikti Toichopoiia)	Greece	Masonry	Masonry/Timber frames	2 or 3	Stone / brick (plaster exterior)	O	Nails and special carpentry

Himis	Anatolia, Turkey	Masonry stone	Timber frames+ masonry	1 or 2	Masonry / adobe/ samdolma/ bagdhadi	O	Nails
Entramados	Madrid, Spain	Masonry	Timber frames+ masonry	2	Adobe / brick	Δ	Notched
Dahjji Dewari	Pakistan	Masonry/ Timber frames	Timber frames+ masonry	1 or 2	Stone	X	Cross-halved, nails, mortise-and-tenon
Dahjji Dewari	India	Masonry/ Timber frames	Timber frames	2 or more	Stone / brick	X	Cross-halved, nails, mortise-and-tenon
Paiantă	Romania	Stone / earth / brick	-	1	Brick/Wattle and daub/earth and straw	O	Cross-halved, nails, mortise-and-tenon
Quincha	Coast, Perú	Adobe	Timber frames	2	Brick, split cane	Δ	Mortise-and-tenon
Gingerbread	Haiti	Masonry	Timber frames+ masonry	2	Brick/ stone	O	Mortise-and-tenon. nails
Kayi pey	Haiti	Masonry / timber frames	Timber frames+ masonry	2	Stone	X	Mortise-and-tenon, nails

(*continued*)

TABLE 2.1 (Continued)

SYSTEM	*LOCATION*	*STRUCTURAL MATERIAL*		*NO. OF FLOORS*	*INFILL TYPE*	*BRACES*	*CONNECTION TYPES*
		GROUND FLOOR	*UPPER STORY*				
ChuanDou	China	Masonry / timber frames	Timber frames+ masonry	2	Brick masonry	X	Penetration joint, Mortise-and-tenon with wedge or dowel
Estaqueo	Paraguay	Timber frames+ stone	-	1	Stone	X	Half-lap or simply suported
Centibera	Nepal	Masonry/ Stone	Timber frames+ masonry	2	Brick masonry/ woven bamboo strips	X	Not clear
Timber framed masonry	Myanmar	Timber frames+ masonry	Timber frames+ masonry	1 or 2	Brick masonry	X	Not clear
Bahareque	Colombia	Earth / Bahareque & timber frames	Bahareque frames	1 or 2	Adobe	O	Mortise-and-tenon, cross-halved

Note: ○ means present; × means not present; Δ means very small occurrence, but present.

REFERENCES

[1] E. Poletti, G. Vasconcelos, and D. V. Oliveira, "Influence of infill on the cyclic behaviour of traditional half-timbered walls," International Conference on Rehabilitation and Restoration of Structures Held in Chennai, India, 13th–16th February 2013, pp. 179–189, 2013.

[2] S. Paz Valencia, "Seismic behavior of traditional timber-framed buildings: Engineered vs. vernacular systems," Master thesis, Minho University, 2020.

[3] E. Comsa, *The Neolitic on the Romanian Territory – Considerations (in Romanian)*. Editura Academiei Republicii Socialiste Romania, 1987.

[4] R. B. Ulrich, *Roman Woodworking*, 1st edition. United States of America, Ann Arbor, Michigan: Sheridan Books, 2007.

[5] E. Tsakanika, "The structural role of timber in masonry of palatial type buildings in the Minoan Crete." Doctor Thesis, in Greek. National Technical University of Athens, 2006.

[6] A. Dutu, "An engineering view on the traditional timber frames with infills in Romania," in *Masonry Construction in Active Seismic Regions*, 1st edition, R. Rupakhety and D. Gautam, Eds., United Kingdom: Elsevier, 2021, pp. 377–420.

[7] Y. D. Aktaş, U. Akyüz, A. Türer, B. Erdil, and N. Ş. Güçhan, "Seismic resistance evaluation of traditional ottoman timber frame Himiş houses: Frame loadings and material tests," *Earthquake Spectra*, vol. 30, no. 4, pp. 1711–1732, 2014. doi: 10.1193/011412EQS011M

[8] N. Quinn, and D. D'Ayala, "In-plane experimental testing on historic quincha walls," SAHC2014 – 9th International Conference on Structural Analysis of Historical Constructions, no. October, pp. 14–17, 2014.

[9] A. Dutu, H. Sakata, Y. Yamazaki, and T. Shindo, "In-plane behavior of timber frames with masonry infills under static cyclic loading," *Journal of Structural Engineering (United States)*, vol. 142, no. 2, 2016. doi: 10.1061/(ASCE)ST.1943-541X.0001405

[10] E. Poletti, "Characterization of the seismic behaviour of traditional timber frame walls experimental results," PhD thesis, Universidade do Minho, 2013.

[11] A. Dutu, J. Gomes Ferreira, and C. S. Dragomir, "Timber framed masonry buildings, an earthquake resistance influenced architecture," in *Structures and Architecture: Concepts, Applications and Challenges – Proceedings of the 2nd International Conference on Structures and Architecture, ICSA 2013*, held in Guimaraes, Portugal, 24th–26th July 2013.

[12] A. Dutu, J. Gomes Ferreira, and A. M. Gonçalves, "The behaviour of timber framed masonry panels in quasi-static cyclic testing," in 9th International Conference on Urban Earthquake Engineering/ 4th Asia Conference on Earthquake Engineering, 2012, pp. 8–13.

[13] A. M. Gonçalves, J. G. Ferreira, L. Guerreiro, and F. Branco, "Experimental characterization of Pombalinos 'Frontal' walls cyclic behaviour," 15th International Conference on Experimental Mechanics, pp. 1–13, 2012.

[14] A. M. Gonçalves, J. G. Ferreira, L. Guerreiro, and F. Branco, "Numerical modelling and experimental characterization of Pombalino 'Frontal' wall," 4th International Conference on Integrity, Reliability and Failure (IRF2013), pp. 1–12, 2013.

[15] J. G. Ferreira, M. J. Teixeira, A. Duţu, F. A. Branco, and A. M. Gonçalves, "Experimental evaluation and numerical modelling of timber-framed walls," *Experimental Techniques*, vol. 38, no. 4, 2014. doi: 10.1111/j.1747-1567.2012.00820.x

[16] S. Stellacci, N. Ruggieri, and V. Rato, "Gaiola vs. Borbone system: A comparison between 18th century Anti-Seismic case studies," *International Journal of Architectural Heritage*, vol. 10, no. 6, pp. 817–828, 2016. doi: 10.1080/15583058.2015.1086840

[17] S. Guerra Pinto, "Numerical Modelling of the seismic behavior of timber-framed structures based on macro-elements," Master thesis, University of Minho, 2017.

[18] L. S. Kouris, H. Meireles, R. Bento, and A. J. Kappos, "Simple and complex modelling of timber-framed masonry walls in Pombalino buildings," *Bulletin of Earthquake Engineering*, vol. 12, no. 4, pp. 1777–1803, 2014. doi: 10.1007/s10518-014-9586-0

[19] S. Paz Valencia, "Seismic behavior of traditional timber-framed buildings: Engineered vs. Vernacular Systems," Master thesis, Minho University, 2020.

[20] F. Parisse, "Numerical modelling of the seismic performance of Romanian traditional timber-framed buildings," Master thesis, University of Minho, 2019.

[21] N. Ruggieri, "La casa antisismica," in *Conservation of Historic Wooden Structures*, G. Tampone, Ed., Florence: Collegio degli Ingegneri della Toscana, 2005, pp. 141–146.

[22] A. Bianco, "La 'casa baraccata,'" *Bio architettura*, Roma, July 2011, pp. 45–49.

[23] E. Vintzileou, A. Zagkotsis, C. Repapis, and C. Zeris, "Seismic behaviour of the historical structural system of the island of Lefkada, Greece," *Construction and Building Materials*, vol. 21, no. 1, pp. 225–236, 2007. doi: 10.1016/j.conbuildmat.2005.04.002

[24] L. A. S. Kouris, and A. J. Kappos, "Detailed and simplified non-linear models for timber-framed masonry structures," *Journal of Cultural Heritage*, vol. 13, no. 1, pp. 47–58, 2012. doi: 10.1016/j.culher.2011.05.009

[25] E. Vintzileou, and P. Touliatos, "Seismic behaviour of the historical structural system of the island of Lefkada, Greece," *WIT Transactions on the Built Environment*, vol. 83, pp. 291–300, 2005. doi: 10.1016/j.conbuildmat.2005.04.002

[26] C. Cancino, S. Farneth, P. Garnier, J. Vargas, and F. Webster, "Estudio de daños a edificaciones históricas de tierra después del terremoto del 15 de

agosto del 2007 en Pisco, Perú," *The Getty Conservation Institute*, pp. 76, 2009. www.getty.edu/conservation/publications_resources/pdf_publications/pdf/damage_assess_esp.pdf

[27] Y. D. Aktaş, "Evaluation of seismic resistance of traditional Ottoman timber frame houses," PhD thesis, Middle East Technical University, 2011.

[28] G. Koca, "Seismic resistance of traditional wooden buildings in Turkey," in *VI Convegno Internazionale*, Messina: Gangemi Editore International, 2018, p. 11.

[29] K. Hicyilmaz, J. Bothara, and M. Stephenson, "Housing Dewari, Dhajji Dewari," in *World Housing Encyclopedia*, R. Langenbach and L. Dominik, Eds., Pakistan: EERI, 2014, p. 34.

[30] Z. Qu, A. Dutu, J. Zhong, and J. Sun, "Seismic damage to masonry-infilled timber houses in the 2013 M7.0 Lushan, China, earthquake," *Earthquake Spectra*, vol. 31, no. 3, 2015. doi: 10.1193/012914EQS023T

[31] A. Dutu, "An engineering view on the traditional timber frames with infills in Romania," in *Masonry Construction in Active Seismic Regions*, 1st edition, R. Rupakhety and D. Gautam, Eds., United Kingdom: Elsevier, 2021, pp. 377–420.

[32] H. Huang, Y. Wu, Z. Li, Z. Sun, and Z. Chen, "Seismic behavior of Chuan-Dou type timber frames," *Engineering Structures*, vol. 167, pp. 725–739, 2018. doi: 10.1016/j.engstruct.2017.10.072

[33] J. Xue, and D. Xu, "Shake table tests on the traditional column-and-tie timber structures," *Engineering Structures*, vol. 175, pp. 847–860, 2018. doi: 10.1016/j.engstruct.2018.08.090

[34] Z. Qu, X. Fu, S. Kishiki, and Y. Cui, "Behavior of masonry infilled Chuandou timber frames subjected to in-plane cyclic loading," *Engineering Structures*, vol. 211, p. 110449, 2020. doi: 10.1016/j.engstruct.2020.110449

[35] A. C. Wijeyewickrema, K. Shakya, D. R. Pant, and M. Maharjan, "*Cuee Report 2011-1 Earthquake reconnaissance survey in Nepal of the magnitude 6.9 Sikkim Earthquake of September 18, 2011*," 2011.

[36] R. Langenbach *et al.*, *Preserving Haiti's Gingerbread Houses: 2010 Earthquake Mission Report*. 2010. New York: World Monuments Fund. www.wmf.org/sites/default/files/article/pdfs/WMF%20Haiti%20Mission%20Report.pdf

[37] T. Hanazato, Y. Tominaga, T. Mikoshiba, and Y. Niitsu, "Shaking Table Test of Full Scale Model of Timber Framed Brick Masonry Walls for Structural Restoration of Tomioka Silk Mill, Registered as a Tentative World Cultural Heritage in Japan" in *Historical Earthquake-Resistant Timber Frames in the Mediterranean Area*, N. Ruggieri and G. Tampone, Eds., Switzerland: Springer, 2015, pp. 83–89.

[38] Y. Endo, and T. Hanazato, "Seismic behaviour of a 20th century heritage structure built of welded tuff masonry and timber frames," *International Journal of Architectural Heritage*, pp. 4–17, 2022. doi: 10.1080/15583058.2022.2113572

[39] K. Yamaguchi, G. Zamani Ahari, R. Naka, and T. Hanazato, "Classification and effects of strengthening and retrofitting techniques for unreinforced

masonry structures," in *The Second Bilateral Iran-o-Japan Seminar on Seismology and Earthquake Engineering of Historical Masonry Buildings*, Tabriz, Iran, 2016.

[40] B. Lizundia *et al.*, *EERI Ecuador 2016 Earthquake Reconnaissance Team Report*, vol. 49, no. 11. 2016.

[41] J. Gutierrez, "Notes on the seismic adequacy of vernacular buildings," in *13th World Conference on Earthquake Engineering Vancouver, B.C., Canada August 1–6, 2004 Paper No. 5011*, 2004, p. 30. doi: 10.5459/bnzsee.38.1.41-49

[42] R. Rupakhety, and D. Gautam, Eds., *Masonry Construction in Active Seismic Regions*, 1st edition. Elsevier, 2021. [Online]. Available: www.elsevier.com/books/masonry-construction-in-active-seismic-regions/rupakhety/978-0-12-821087-1

[43] S. Htwe Zaw, T. Ornthammarath, and N. Poovarodom, "Seismic reconnaissance and observed damage after the Mw 6.8, 24 August 2016 Chauk (Central Myanmar) Earthquake," *Journal of Earthquake Engineering*, vol. 23, no. 2, pp. 284–304, 2019. doi: 10.1080/13632469.2017.1323050

3 Romanian Traditional Architecture

Traditional Romanian architecture for residential houses in rural areas is mainly based on timber structures, having different types of layouts depending on the location (mountain, fields, etc.) and availability of the raw material. Unreinforced masonry, adobe, and cob houses can also be found within the Romanian territory, but we mainly focus on the ones with timber frames due to the resilience they showed over time in both earthquakes and durability.

The spreading of traditional house types within Romanian territory was studied and published in several documents [1] and [2] and several maps were issued depending on the type of house and the construction materials. In the latter one, a map has the exterior structure of the wall, specific to each traditional house, which in fact defines the type of house.

The RURAL Group of OAR (Architects Order of Romania) has conducted an impressive and recent work, focused on the volumetric and architectural details, and on the identification of the characteristics of traditional houses [3]. Also, URBAN-INCERC Iasi has conducted several research projects on compatible materials for traditional houses, using the local and natural available ones, and this research focused on the thermal insulation properties of such materials [4, 5]. Another group of researchers from URBAN-INCERC Cluj Napoca focused on the material properties (both thermal and mechanical resistance) of adobe bricks, mortar, and plaster with different compositions (i.e. reinforced with hemp fibers) [6, 7].

A description of general traditional features is provided in [8], for a general understanding of the types used in Romania. Further, elements of the Romanian traditional houses with timber frames and various infills are summarized with just enough details to have a bird's eye view of what to look at, when an engineer is in front of such a house.

DOI:10.1201/9781003405375-3

3.1 TRADITIONAL *PAIANTA* HOUSE TYPOLOGIES

A type of traditional timber house is one with posts and beams and various infills. The appearance of this type in Romania, some say, was inspired by the *fachwerk* houses in Germany [9], or, more likely, as an adaptation to the decrease in the availability of timber. Instead of massive logs, beams and posts with smaller cross-sections were used, initially infilled also with timber logs, but where these were even harder to find (in the plain areas, for example) other infill materials were used, such as mud, straw, wattles, etc.

In my studies, I considered that those are also part of the "*paianta*" typology. This book focuses only on braced timber frames with infills, given that they have proven their resilience almost all over the world [10] and that the details that helped them withstand seismic events should be well understood and transferred to future generations.

Based on several field investigations, the following types of structural systems that may be included in the *paianta* typology were identified, as explained more thoroughly in [9]:

- with timber skeleton and brick masonry infill structure (Type 1 – "*paianta*" – classic) (Figure 3.1a);
- with timber skeleton and wattle and daub infill (Type 2) (Figure 3.1b and c);
- with timber skeleton and horizontal strips, infilled with earth and straw (Type 3, Figure 3.1d);
- with timber skeleton, infilled with horizontal timber logs (Type 4, Figure 3.1e);
- with timber skeleton and AAC (autoclaved aerated concrete) masonry infill (Type 5, (Figure 3.1f). This type is actually a deviation from the traditional house, and further description will not be detailed.

The figures show houses in a rather poor state just to present the structural system, but many of them, which are representative of the local culture, are well preserved and are covered with plaster and finishing.

Except for the last type, which is a modern and wrong interpretation of the classic "*paianta* ", they all may be considered within the *paianta* typology, having the same timber skeleton, but different infills.

FIGURE 3.1 Structural types of *paianta* houses: (a) Type 1 – timber skeleton and brick masonry infill structure; (b) Type 2 – timber skeleton and wattle and daub (horizontal layout); (c) Type 2 – House with timber skeleton and wattle and daub infill (vertical layout); (d) Type 3 – timber frame and earth with straw infill structure; (e) Type 4 – timber skeleton and horizontal timber logs; (f) Type 5 – timber skeleton and AAC (autoclaved aerated concrete) masonry infill.

3.2 HOUSES' GENERAL CONFORMATION

In Figure 3.2 one classic type of *paianta* is shown, with all the components which are further explained. Thus, they are composed of a main frame (skeleton), having posts, beams, diagonal braces, and infills. Foundations are usually made of stone and raised from the ground.

Most of the houses are one story high, but occasionally, they can be with an upper story, having the architecture as a house with the living space on the upper floor, while on the ground floor, the owners have the storage rooms.

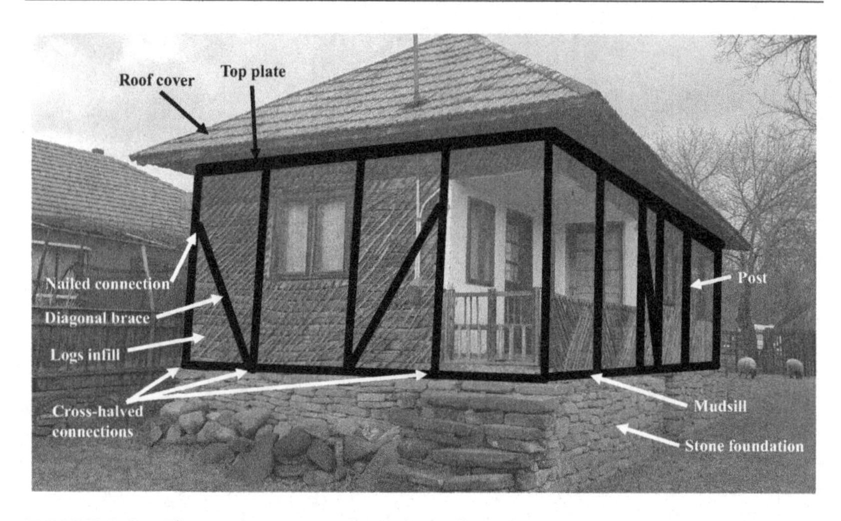

FIGURE 3.2 The components of a *paianta* house.

3.3 COMPONENTS AND HOW TO BUILD A *PAIANTA* HOUSE

3.3.1 Material and Tool Preparation

We live now in a world where things must happen fast, and people tend to want things even faster. But in history, people would take their time to properly prepare the construction materials. They first gather them, then arrange them in a covered place, and then wait for them to dry or settle for some time, sometimes years. And only after adequate preparation, the materials could be used for the construction.

The *timber* was cut when it was cold enough outside, and the sap content was at a minimum so that it could be exploited more easily. In some areas, the tradition was to cut the timber when it was a new moon, not a full moon [11]. The carpenters knew, from generation to generation, how to identify the good timber, when to cut it, and how to dry it. For example, [9] states that the trees on the top of the hills, the most exposed to wind, growing slower than the others, had the densest fibers and if the cut is made in the summer, the timber cracks enhancing the biological decay. After the cutting of the tree, the first operation would be the removal of the branches and the small protruding parts. Using the axe, the bark was then removed. By this operation, the drying would become faster, and rooting was prevented. The timber thus prepared was stacked sitting on some supports, to be raised from the ground and thus protected from humidity, and covered, if possible, in a shadowed place [11]. It would stay like that for more than a year before it would be used for construction.

Timber nails or dowels were also prepared, made of yew tree, chopped with a small axe into 25–30 mm diameter and 250–300 mm long. In the same way, the *shingle* was prepared, from secular firs with a straight fiber.

Unprocessed *stone*, usually from the river, was mainly used for the foundations. It was also collected earlier and stacked in piles near the future construction site. It was left there waiting until one winter would pass, thus, due to the freeze-thaw phenomenon, the improper stones would be destroyed, and the proper ones could be used in the construction [9]. Sometimes, if in the area there was a skilled master, the stone would be processed. But with the openings of stone quarries, this type of skill disappeared, as it became unnecessary.

Gravel was also collected from the riverbanks, which should have been clean, uncontaminated by mud, and have a granularity of 4–71 mm. It was staked in compact piles, so as not to mix with other construction materials [9]. The same is valid for the sand, and it has a granularity of 0–4 mm. As a check, if the sand is proper for the construction works when one closes the fist full of sand and opens it, the hand should not remain dirty.

A long time ago, the carpenters didn't use many *tools*, as the iron was expensive and hard to find. The number of tools depended on the skills of the carpenter and also on the local architectural style. The most popular tool was the axe. Then, the straight draw shave tool ("*cutitoaia*") was also used, especially for the shingle, to smooth its two faces and with a woodworking planner ("*rindeaua*") the two margins were straightened. The drill was used for the joints, since the timber dowels were mainly used. For the old (19 century) houses, the chisel was not used that much. The few tools mentioned above were made by blacksmiths, sometimes by Roma people, which also introduced the "gypsy nails" made of iron.

The erection steps of a *paianta* house follow a similar procedure as in the case of log houses, which is actually the oldest and most widely spread construction type, at least in Romania. It is more specific to the forest and hilly mountain areas, where access to the timber was easy. In the other regions, the traditional houses had to *adapt to the available materials*, so less and less timber was used, with a smaller cross-section, and other materials were included in the house to replace the hard-to-get timber. This is the case for straw and clay *infills*, wattle and daub, mud bricks, and more recently, burned clay bricks.

The *clay* was usually collected from the "clay hole of the village", from where all the villagers were digging and using the earth. Of course, the mechanical properties of the earth differ from area to area, and while in some areas the clay is more cohesive, in others, it is pretty plastic in its behavior after drying. Local people knew this, and depending on the area, the infill solution was selected as a local culture, also based on the knowledge of the better type corresponding to their available materials. It should be as yellow and sticky as possible [9]. Some natural additives can be used to improve the properties of clay. One of them is the casein, which is produced from sweet cow cheese, and mixed with slacked lime paste (1:1), then filtered and becomes a white viscous fluid, adding a stabilization increase to the mix. When mixed with horse dung, the mix becomes a good waterproofing material.

The *wattles* for the infill were usually made of hazelnut trees and simply collected and stored. One thing to consider is whether they are placed at their full length, interlaced around a post (the intermediary one in a wall), or cut into shorter pieces and inserted between two posts. After the branches are cut from the tree, while they are still green, they are straightened and stored. The more they become dry, the more they lose their flexibility, and cannot be easily straightened.

The *bricks* were either made of mud, erected in forms, left to dry, and then burned in the oven or, for more recent applications, they are industrial burned clay bricks, made in the factory. The mud (clay) for the bricks is carefully selected, usually in autumn, stacked in piles, and left outside to experience freeze-thaw and dry-moist cycles. After they are burned in the oven, the bricks are left to dry, in covered and well-ventilated barns. When the bricks do not decrease in weight, then the drying process is finished.

The *lime* is recommended by [9] to be manually slacked, and it is thought to be more resistant than the non-slacked lime. In order to slack it manually, a lime pit is used (like a box) and made of timber planks, resulting in a final rectangular shape with dimensions of $2 \times 1.5 \times 0.3$ m. On one side an opening is left, covered with a wire mesh (maximum grid hole of 3 mm) and also provided with a door through which the lime paste is drained into a lime hole. The bottom of the lime pit is made to be waterproof and with an inclination towards the lime hole. The lime hole is dug, with attention to the cohesion of the soil, so that the walls of the excavation do not collapse. To slack the lime X in the lime pit, water is poured until ¼ of its volume and then the lime balls are introduced using a shovel. Then, more water is poured until the total quantity becomes 2.5–3 times the lime volume in the lime pit, and 1 liter of gas or petrol is also inserted into the pit. The mix is continuously mixed until the consistency increases and becomes a uniform milky paste. After the boiling stops, the little door is lifted, and the milky paste drains through the wire mesh opening. The water goes into the soil, while the milky paste remains. It can be used for masonry joints after two weeks of storing it in the pit, and to be used for plastering, it should be kept for at least two months in the pit. To prevent the excessive drying of the paste while storing it, a 15 cm layer of sand is placed on top of it [9].

The *clay mortar* must have a 13–15 cm consistency [9], tested with the slump cone. When it is excavated, the first layer of the soil (containing vegetation) is removed, and the actual clay is piled and strongly watered. It is kept wet for a few days, and after that is put in a pit, built also from timber planks, and more water is added. The content is mixed with a hoe, until the mix is homogenous and with the required consistency. If a higher strength is needed for the mortar (higher than M10), the resulting mix is passed through a mesh with a grid of 3 mm, and it can be used as is. The remaining clay is put back into the pit and mixed with new clay, and the same steps are followed. The mixture can also be done in a usual mixer for mortars.

The *lime mortar* is also done in a plank pit, where the lime paste with a consistency of 12 cm is laid, according to the recipe. Water is added, as much as needed to turn the mix into a viscous milk, being homogenized with a hoe. Sand is then added, and the content is mixed until the mix becomes homogenous. Water is added until the necessary consistency is reached [9]. Lime mixed with oakum (laid in a lime pit in nine rows, and then mixed carefully for a homogenous dispersion), was sometimes used as interior finishing, and it was thought to be a very good method for thermal insulation.

3.3.2 The Foundations

The existing traditional houses have foundations if they are newer than 1944 [12]. In some situations, the posts were directly buried into the ground, or a trench was dug and then filled with a mortar, called "*ciamur*", made of yellow soil (clay) mixed with straw or chaff (called "*pleavă*"). If the groundwater level was high (less than 50 cm below the ground surface), no trench was executed [12]. In some areas (i.e. Prahova County), a 50 cm-deep trench was dug and filled with gravel and the foundation was raised above the ground by about 50–150 cm, maybe due to flooding hazard. A slacked lime was poured on the gravel in the trench, whenever possible. Then, stones were cut to complete the foundation. The trench was as deep as between 50 and 80 cm, usually until reaching the bedrock. Sometimes the trench was filled with rammed earth and had a width of 80 cm. Between the foundation trenches, rammed earth was placed. Some houses, in particular those of wealthier people, had cellars underneath some of the rooms [9].

The foundation component materials, depending on the local condition (i.e. region, resources, tradition/cultural identity), may be river, quarry or carved stones; fired or just dried in the sun clay brick masonry; timber piles;

combined oak timber with layers of quarry stone; concrete with large rocks (with coarse aggregates); adobe; grassy soil; concrete or concrete ballast.

In many situations, the foundations are "dry", i.e. having no binder; this is the case when they are made of river stone, and this has an amazing property: canceling the capillary action of water, and thus not reaching the mudsill. However, if a binder exists, it can be made of earth (soil) mixed with dung, straw or wheat chaff ("*pleavă*") or mortar made of lime, sand or earth, sometimes with gravel (lime: sand proportion 1:3), sometimes with cement (more recently) [9].

The exterior finishing of the foundation, if present, was made of earth ("*pâmânt*") mixed with cow dung and sand, sometimes with clay ("*lut*") and soot ("*funingine*") [12]. Recently, a cement paste is also used.

When the mudsills are supported on *big stones* (or oak tree logs), at the *corners of the house*, the space between these was infilled with rammed earth, and then sand was applied at the interior surface, while on the outside clay was used to plaster them. These aimed to give a certain thermal comfort in the winter season, keeping the cold from entering the house. If the foundation was made of dry-stone masonry (no mortar), a special focus was given to the interior earth infill. This was the case for very old houses, maybe in the mid-20th century. Recently, traditional houses have foundations buried in the ground, especially where concrete is used. Underneath the mudsills, after joining them at the corners, between the corner stones, sometimes intermediary supports are used, and between them, other river stones are erected, to make a compact foundation.

Nowadays, the building standards recommend that the foundations are made of concrete and connected between them, to form an assembly that works together in case of seismic actions [13]. Most often reinforced concrete beams are used, resting on unreinforced concrete blocks at the intersections of the walls, which in turn rest on the compacted soil and a sand layer of 5 cm. After the concrete footings and beams are cast, the spaces between them are backfilled and compacted well. On top of it, a 10 cm layer of gravel is placed to block the capillary action through the soil. Then, a polyethylene sheet is placed with thermal insulation on top, which can be extruded polystyrene of a minimum of 5 cm. Then, on top of the beams, with reinforcement connected and protruding from them to create a strong diaphragm, the floor slab is cast also made of concrete with the same class as the one for the beams, minimum C20/25 class.

3.3.3 The Mudsills (If Existing) Erection

According to [12] the *mudsill* (present only in post-1944 houses) was made of round or carved timber, usually hardwood, but in fact, any available species

FIGURE 3.3 Mudsills types of a *paianta* house.

could have been used: fir, beech, acacia, willow, oak, alder, hornbeam, ash, birch, etc. Mudsills could have been placed either directly on the ground (Figure 3.3a), or on pilasters (called "*chei*" or "*lespezi*") of 40–50 cm, made of stone, clay/stone brick masonry or concrete (since 1930), placed only at the corners of the house (Figure 3.3b, c), on small piles (called "*pari, gâşte, butuci, bulumaci, căsuţe, chituci, tufani*", depending on the area) inserted into the soil about 90 cm deep (or not) and connected to the mudsill with 20 cm-long nails (Figure 3.3d), or on the river (Figure 3.3e, g) or quarry stone (Figure 3.3f), burned clay brick masonry (Figure 3.3h) or concrete foundation (after 1916; in general, post-WWI) (Figure 3.3i).

More than 200 years ago, the mudsills were thick fir trunks (30–40 cm in diameter), carved on one side or on two sides. In very old houses, they were even left round. The mudsills are two types: the longitudinal ones and the transversal ones, sometimes different in thickness. When the timber was easy to get, the mudsills were as long as the house, and later, or in other areas, they were connected by nailed half-lap joints, sometimes with clamps instead of nails. The corner joints are also usually *half-lap joints, but they can also be dovetail joints, and they can be also fastened with additional clamps*. Then, the transverse mudsills are fixed on the longitudinal ones, also by half-lap joints. More recently, the cross-section would not be over 20×20 cm, and usually 15×1×5 cm.

To understand the layout of the mudsills, and how they rest on a stone foundation, Figure 3.4 shows how the mudsills' grid (longitudinal and transversal)

FIGURE 3.4 Mudsills grid, sitting on a stone foundation.

rest on the stone foundations. Between the stone foundations, gravel or rammed earth is infilled.

The *timber species* in the old houses would usually be hardwood (oak tree or acacia), but more recently, since it was not easily available, softwood (fir) was more popular.

If the foundation is made of concrete beams resting on blocks, the mudsill should be carefully supported, and no cardboard should be placed between the concrete and mudsill (Figure 3.5c), because the capillary action of water through the concrete brings the moist into the cardboard and this keeps it right underneath the mudsill which degrades in time due to enhancement of the biological attacks. A solution would be to add a bituminous membrane between the mudsill and the concrete (Figure 3.5b). Although dry-stone masonry placed beneath the mudsill is an excellent solution to prevent capillary water from reaching the mudsill, from the seismic behavior point of view, it is not a safe and predictable solution. The connections between the foundation and mudsill can be with hold downs, protruding from the concrete where they were placed during concrete casting and nailed to the timber posts with minimum three screw nails. Figure 3.5, shows a type of hold-down tested in Japan, and the satisfying experimental results in comparison with other types of traditional Japanese and cross-halved (half-lap) connections are detailed in [14].

FIGURE 3.5 Connection types between the concrete foundation and the mudsill: (a) hold-down connection, recommended; (b) metal plates embedded into the concrete foundation which is later screwed into the mudsill, and bituminous membrane between them, recommended; (c) no connection to the concrete foundation, and cardboard between them, not recommended.

3.3.4 The Timber Framing

The most often used *timber species* were oak and holm, but fir and spruce, or a mix of coniferous wood can also be used [9]. Nowadays the timber is mainly fir, with a lower strength than it used to have.

The *timber frame* structure is formed with vertical (posts) and horizontal elements (mudsills and top plates) and bracings ("*paiante*"), which are positioned at the corners and intersections of the walls, but not always in a coherent and symmetric distribution. The cross-section of the elements is a maximum of 12x2 cm. Bracings are usually stopped before the upper post-top plate connection, this being a curious detail and it is thought that it was done for the economy of the material or maybe the bracings initially have a constructive role, just to support the posts until they are connected with the beams and the whole structure can stand by itself. This detail's influence on the lateral strength of the wall will be further explained in Chapter 5, Section 2.

The *joints* are often mortise-and-tenon type (Figure 3.6a), or cross-halved (half-lapped) (Figure 3.6b), or simply nailed (Figure 3.6c–e). Occasionally, steel clamps (Figure 3.6f) are added to increase the resistance and stiffness of the joint [15]. In some situations, the mortise-and-tenon type connections have additional nails (Figure 3.6a). Generally, the connections between the

FIGURE 3.6 Connection types of a *paianta* house: (a) mortise-and-tenon with nails; (b) cross-halved; (c) overlapping and nailed joint between the post and the mudsill; (d) notched connection between the brace and the post; (e) overlapping and nailed joint between the post and the brace; (f) overlapping and nailed joint between the post and the mudsills (longitudinal and transversal); (g) clamped.

timber elements are not carefully executed, since they were made by axe, and many gaps can be observed. Nails can have an influence on the ductility of the timber, but in many cases, they are lacking.

The diagonal braces are usually nailed or clamped to the posts and mudsill, but some other types of lapping are possible, sometimes (for old and authentic houses) with a timber dowel. To mark the window and door openings, posts are laid on each side, and the openings are just in between them. The mortises in the mudsills and top plates are usually made with a chisel and hammer, and they cross the entire thickness of the timber element. Then, the tenon of the column may be initially chopped with an axe and finished also with the chisel. The timber diagonals for the *paianta* houses possess a specific feature as they are connected to the post, not in the corner connection with the top plate but lower, almost 30 cm below the upper corner joint. This may be due to economic reasons, as the diagonal elements would not need to be too long. But, it is interesting to know that the experimental tests conducted on walls with diagonals connected directly in the upper corned joint gave very similar results in terms of maximum force, although lower stiffness, with the same type of wall having the authentic joint of the diagonal brace to the lost lower than the upper corner. For the latter specimen, also less damage was observed during the test for the bottom connections (post-top plate). More information on this is thoroughly described in [16] and later in Chapter 5.

FIGURE 3.7 Tools and methods to create a mortise: (a) chain saw; (b) sward saw; (c) drilling machine; (d) chisels set; (e) whetstone; (f) carpentry hammer; (g) hammer; (i) tape measure; (h) drill holes to start chiseling; (j) chiseling into the mudsill until the right depth; (k) final mortise.

The *top plate* is erected too, in this step, tying the posts and also having the role of supporting the roof framework. The cross-section is similar to the posts.

The erection process is shown further, for the timber frames of *paianta* walls using modern tools and nowadays available materials. A mortise-making and necessary tools are presented in Figure 3.7. Initially, the timber's moisture content was measured, and it should not exceed 14%. The posts' tenons (Figure 3.8) are carved using a circular chainsaw, chisel, and hammer, and for a post, it should take less than one hour if the timber has no knots and max. 1.5 hours if the worker is medium-skilled (engineer, but not a specialist in traditional construction techniques). Then, after marking the mudsills with the mortise position, small chainsaws can be used to make the rough cut, but in the end, the chisel and hammer would make the necessary precise mortise. Usually, depending on the skills of the carpenter, gaps may exist between the tenon and mortise. To create the tenons of the posts, a fixed vertical chainsaw machine can also be used, to have a precise cut, but any other chainsaw tool can be used for this.

After joining the frame components, nails or clamps can be hammered into the mortise-and-tenon connections for additional stiffness.

Nowadays, metal screw nails can also be used, due to the easiness of the workmanship, just using an electric drilling machine.

If cross-halved connections are chosen, they can be executed as shown in Figure 3.9.

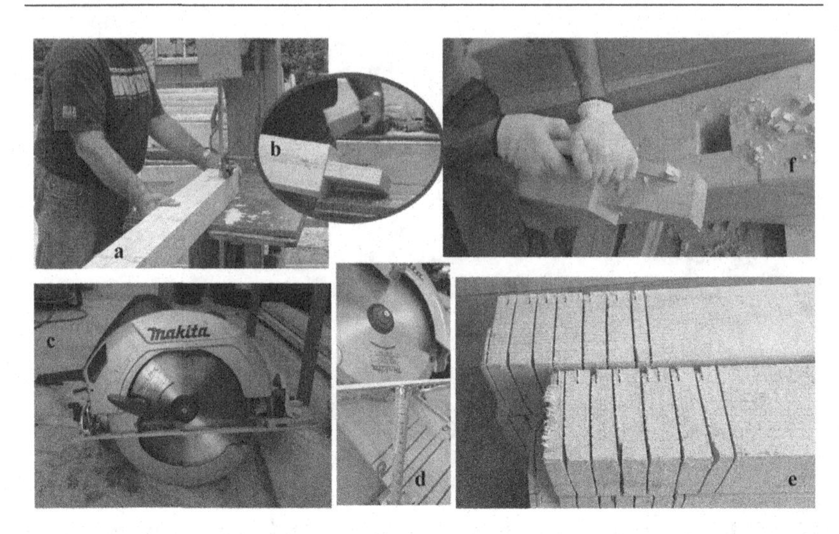

FIGURE 3.8 Tools and methods to create a tenon: (a) band saw; (b) resulting tenon from band saw; (c) pendulum saw; (d) setting of the target depth of the pendulum saw; (e) making several cuts at the same depth to create the tenon; (f) removing the cut wood and finishing the tenon with the chisel.

FIGURE 3.9 Tools and steps to create a cross-halved connection: (a) mark the dimensions on the beam; (b) make cuts with the table saw until the target depth (half of the cross-section); (c) result after making the cuts at the target depth; (d) using chisel and hammer remove the timber layers after the cut; (e) measure the final depth and (f) make the other cross-halved part (post) and connect them using a timber hammer; (g) final connection; (f) wood hammers.

3.3.5 The Infills

The *infills* may be made of earthen bricks (fired) (Figure 3.1a) or wattle and daub (Figure 3.1b, c), or horizontal logs (Figure 3.1d). In rare situations, one can find also a combination of several types of *paianta* infills, usually due to reparations of the original type after earthquakes and the newest method is with AAC bricks (Figure 3.1e).

The mortar for the infill may be made with clay, chaff, and minced straws, mixed together by squeezing them with the feet for about six hours, preparing in one batch only the necessary quantity to be used during a day of work [9]. In the case of using earth mixed with straw as an infill, after placing the first layer all around the house walls, this would be left to dry for a week, then the second layer would be placed, and so on.

The infill preparation steps depend on the infill type. Thus, for *type 1* (as defined in Chapter 3, Section 1) with the fired earth bricks infill, the execution steps are shown in Figure 3.10 and are as follows:

FIGURE 3.10 Making of mud bricks and mud mortar, then using them for the construction of a type 1 *paianta* wall: (a) collecting the clay; (b) preparing the bricks' molds; (c) drying the bricks in the sun after forming them in the mold'; (d) firing the bricks in the oven; (e) delivering the bricks and stockpiling them near construction site; (f) mix the mud mortar using an electric drill with a mixing paddle; (g) or mix the mortar using a hand-held paddle mixer for the mud mortar; (h) a bucket of water; (i) bucket dipper; (j) round head trowel; (k) the consistency is appropriate when the trowel can stay vertical into the mud mortar; (l) applying the mortar on the brick; (m) fixing the brick into the wall; (n) checking the horizontality of the bricks layer with a bubble level; (o) adjusting the horizontality by soft timber hammer hits.

- The earth is freely collected from the village's "earth hole";
- The mud bricks are formed in small rectangular timber molds and left to dry;
- The dried bricks are burned in the oven;
- The bricks are left to settle and then can be used for construction;
- The mortar is made of clay, water, and sand brought to a consistency strong enough to not flow from the trowel when inclined, but soft enough to be able to easily fix the brick in the earth layer with soft timber hammer hits;
- The masonry is laid with mortar joints of 1 cm thickness;
- The infill is left to dry, and thus it cracks, for at least one month, and then an additional layer of clay is applied in a thin layer, combined with straws.

For *type 2* of *paianta*, with the wattle and daub infill (Figure 3.12), the steps are:

- Mounting of the intermediate posts, having a cross-section half the size of the posts;
- Mounting the support timber strips (20×50 mm) on the interior of the posts to which the branches are nailed;
- Cutting the branches to the appropriate size with pliers;
- Interlacing the wattles (branches) through the timber frame's posts and intermediate posts, nailing them to the support timber strips;
- Preparing the mud mortar, which includes straw, besides the clay and the water. Traditionally the mixing is done by foot (Figure 3.11), and should take up to six hours for a correct mix, but nowadays it can be done efficiently with a hand paddle mixer, too;
- Application of the mud mortar on the wattle and daub infill;
- Smooth the surface of the wall with a self-leveling trowel or by hand;
- The infill is left to dry, and thus it naturally cracks, for at least one month, and then a layer of clay is applied including also straws.

To build a *type 3* infilled *paianta* wall the following steps should be taken:

- 20×50 mm timber strips are nailed to the timber frame, on both sides, offset, thus forming a formwork structure for the mortar to be applied between them;
- The mud mortar is prepared with water and chopped straws (Figure 3.11a–c);
- The balls created from mud and straw are put in a wheelbarrow and then laid in the space between the strips and the timber frame;
- The infill is left to dry, and thus crack, for one month, and then a layer of clay is applied including also straws.

FIGURE 3.11 Execution of a type 2 *paianta* wall: (a) Collecting the branches by cutting them with pliers, (b) Delivering them to the site; (c) mounting timber sticks (20x50 mm cross-section) on the interior face of the posts and braces/diagonals; (d) interlacing the branches between two posts and an intermediary timber stick (20x50 mm cross-section); (e) and (f) executing as in Figure 3.12 and applying the mud mortar mixed with straws on the laid branches, with care that the mortar goes from one side to the other; (g) a trowel is used to finish the wall; (h) the surface of the mortar is leveled on both sides.

For the *type 4* of *paianta*, with the timber logs infill, the following steps are necessary (Figure 3.13):

- In some situations, carve the interior of the columns and diagonal braces with a groove 2 mm thicker then the end of the horizontal log or create it by applying 20×20 mm strips on the insides of the columns, such that in between them the logs will be fixed tightly;
- To fill in the empty space between the timber frame and the diagonal braces timber elements (logs) are used, laid horizontally. The timber logs are processed on all four sides, thus they are not exactly as the round logs found in the classic log houses;
- The cross-section of the logs should be the same as for the posts and braces;
- Cut the timber elements at the correct length, wither by pendulum saw, chain saw, or table saw to fit within the main timber frame;
- If necessary, carving of the tenons, at both ends of the logs, which will be inserted in the grooves carved on the posts and braces;
- Depending on the skills of the workmanship, carving can be replaced by just overlapping the logs on the posts and braces, and nailing them;

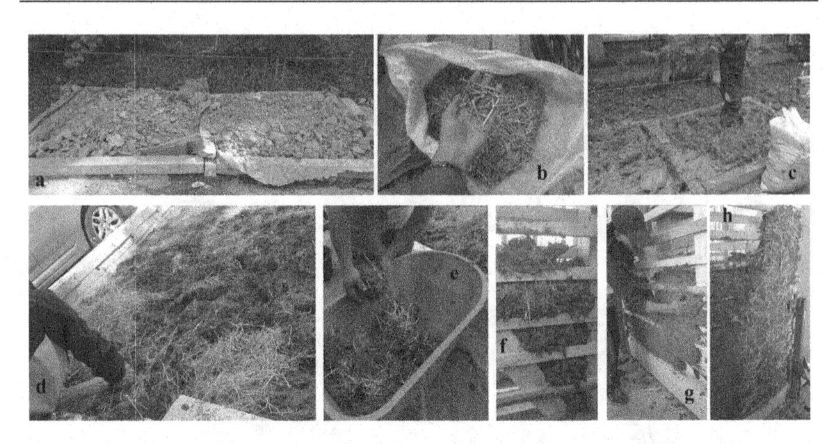

FIGURE 3.12 Preparing and applying the mud mortar with water and chopped straw for type 2 and type 3 *paianta* walls: (a) preparing a place, surrounded by timber beams/planks, into which the clay is laid on a plastic sheet; (b) preparing the straw chopped with an axe; (c) mixing the clay with the water and straws, and stepping on it by foot, wearing rubber boots; (d) the straws can be cut also by hand, by rubbing them into each other; (e) creating clay balls with straws; (f) inserting the balls within the timber frame; (g) leveling by hand the exterior of the wall; applying straw to avoid significant cracking.

FIGURE 3.13 Erecting the type 4 of *paianta* infill: (a) marking the positions of the logs on the main timber frame and measuring the dimensions of each log; (b) cutting the logs with the table saw; (c) inserting the cut logs into the main timber frame avoiding gaps between them; nailing the logs to the main timber frame in an inclined position.

- The erection of the horizontal timber logs should start with the short (length) pieces, so they could fit between the posts or the post and brace;
- At the end, the walls can be left like that or plastered with a layer of clay between the logs.

3.3.6 The Roof Framework Erection

The *roof* framing can have many shapes and materials. It strongly depends on the location of the house (mountain, field, etc.), the dimensions of the house, the workmanship, and the materials available [9]. Thus, the rafter ends may be supported either on the top plate of the wall, i.e., on the perimeter of the building, or they can extend as an overhang, covering the porch as well. The extension of the eaves is very important in the protection of the house against rainwater. The roof framework supporting the rafters may be made of a single ridge beam, sitting on props, which, in turn, sits on a bottom beam. The transversal beams called "*cordiţe*", if existing, are usually spaced 60–80 cm apart, projecting on the exterior of the roof, as overhangs, by about 60 cm [9]. The connections can be made cross-halved, but more often with iron/steel nails or clamps.

The shape may be of shed ("*într-o apă*"), gable ("*în două ape*") or hip ("*în patru ape*") type (Figure 3.14). Ventilation gaps are sometimes provided on the roofs which also serve to illuminate the attic.

The usual shape of the *paianta* roof is the hip roof, especially when located in a strong wind and heavy rain area. The height of the roof varies with the

FIGURE 3.14 Roof types of a *paianta* house: (a) hip; (b) gable; (c) roof with ventilation gaps.

location. But as a reference, in plain areas the height is 2/3 of one of the walls, in a hilly region the height is the same as one of the walls, and in mountain areas, it can be higher [11].

First, on the posts of the walls, supported by braces to remain vertical, the top plates are erected. On the top plates, joists are placed, to support the ceiling's decking which in turn is made of thick planks, made of fir wood, in old times carved with a small axe. The joists are connected usually with notched or cross-halved joints, to the top plates, perpendicular to the joists. To seal the gaps between the planks, in old times clay was used, applied from above, through the attic, or another layer of planks was applied. Sometimes the clay would be applied on the bottom side, to the ceiling of the room it is covering.

On the top plates, the rafters are sitting, and forming an angle to give the shape of the ridge of the house. Traditionally, the connection with the top plate is cross-halved, so the top plate is carved with half of the cross-section of the rafter, and the rafter would be carved with half of the cross-section of the top plate, and a timber dowel connects tightly both. But since the appearance of the nails and clamps, they are often used to connect rafters with top plate, although it is known to produce the faster degradation of the timber due to the rotting of the metal fastener, and also the splitting of the timber due to the thickness of the clamps.

The rafters in the longitudinal direction of the house are mounted in pairs, creating an angle at the ridge, and then the ones in the transverse direction are erected.

Starting from the lower part, depending on the shingle's length, at every 40–60 cm horizontal timber strips are nailed to the rafters, in old times, traditionally using timber dowels. On these, the shingles will be mounted. Although these dowels were originally made of yew trees, since their leaves are poisonous to animals, overuse has reduced the availability of this species. But more recently, iron nails started to be used, made by blacksmiths, or, even more recently, industrial nails appeared and were also used.

3.3.7 Roof Cover Application

The *roof cover* types are also diverse, depending again on the location of the house, materials availability, and workmanship skills. Thus, where available, reeds can be used (i.e. in the southeastern part of Romania – Dobrogea), but also earth or pantile ("*olană*") (Figure 3.15a). Straws may also be used (Figure 3.15e), more likely from wheat or rye ("*secară*") straws. To fixedly attach the straws to the roof structure some strips are used during the first years or tied with twigs or wires. The roof cover can also be made with corn

FIGURE 3.15 Roof cover types of a *paianta* house: (a) pantile; (b) shingle; (c) ceramic tiles; (d) metal sheet; (e) reed/straw; (f) asbestos.

stalks or long planks. The shingle ("*şindrilă*") (Figure 3.15b), sieve ("*şiţă*"), and clapboard ("*draniţă*") are also an option, and differ according to their dimensions: the sieve is small, the shingle is average size and the clapboard is large [9]. These latter three can be made of fir or beech, sometimes of oak. Tiles ("*ţiglă*") were also often used (Figure 3.15c), as well as galvanized metal sheets (Figure 3.15d). Later, asbestos-cement sheets ("*azbociment*") (Figure 3.15f) were used [9].

The shingle is one of the most popular traditional roof covers. The fir trees, having straight fibers, are cut in logs with a length of 80–100 cm, then, using the axe and a hammer, the shingles are chopped from them in 1 cm thickness. After chopping, they are stacked and left to dry. Before being used, they are smoothed with the draw shave tool ("*cutitoaia*"), and their margins are profiled with a timber jointer.

3.3.8 Plastering of the Finished House

The wall *envelope* [12] may be made either of timber wainscot ("*lambriu*") (Figure 3.16a and b), or of three layers of plaster made of dung and earth (2:1) (Figure 3.16c). The *finishing* may be also based on lime (Figure 3.16d), dung, earth, straw, sand, and, more recently, cement, in different recipes. Sometimes brick or tile debris can be mixed in the finishing as well. The finishing is applied directly on the walls or on a support layer. The support timber layer may be chopped with the axe, to enhance the connection of the plaster to the

FIGURE 3.16 Wall envelope types of a *paianta* house: (a) short wainscot; (b) long wainscot; (c) interior lime finishing; (d) earth and sand exterior finishing, covered with a layer of lime; (e) lime and sand exterior finishing, covered with a layer of cement plaster.

FIGURE 3.17 Support layer for the plaster: with obliques timber strips (left) and wire and nails (right).

support, or of timber strips (Figure 3.16a), twigs laid in zigzag, wire mesh, and nails [9] (Figure 3.16b).

Old traditional houses used to remain unplastered on the outside. But more recent houses are also plastered with clay and lime on the exterior. The clay was mixed with chopped straw. After the mixture was homogenized, it was moved into the house with a fork and from there, by hand, applied on the walls. At the end, it was smoothed with a piece of plank. For increased adherence, a popular technique was to stick into the walls (on the horizontal logs of the type 4, for example) timber dowels, on which the clay layer was applied. Then, even more recently, instead of the timber dowels, oblique fir sticks are nailed to the wall, and in the case of the ceiling, they were used in two layers to strongly support the clay plaster.

Plastering of the house was a community effort in the old times, where many villagers and kin would come to help, and once finished, a big party was thrown.

Due to the nature of the clay, the plaster would dry and crack, and then, another layer was applied, this time with a yellow and sandy clay mixed with chaff and water, and then, a second layer of finishing was applied, a mix of clay with horse dung, and in the end, all the walls are painted with lime. As described in [11], this technique dates back to the first centuries after Christ and was spread all over the country.

The new trend is to also add, before the plastering, a layer of thermal insulation. Although the most common one is with polystyrene panels, it is not compatible with earthen materials, and thus, natural materials insulation panels, such as compacted straw (wheat), or reed are recommended to allow the "breathing" of the wall materials, so humidity will not stay inside the walls, such as in the case of polystyrene.

3.3.9 Floor Decking

The *floors* are made of wood planks (the ones above the foundations) and the ones above the ground floor are usually made of timber joists with wood planks, or wattle and daub (Figure 3.18b), or sometimes reed. Above the main structure of the floor, there is a layer of earth (mud) on top of the planks. Between the wattles, if existing, a lime and oakum mix can be infilled, having an insulation role also. The planks are connected to the timber beams by nails, which gives some stiffness to the floor in-plane, but still not enough to be considered a rigid diaphragm.

At the old houses, the floors were made of rammed earth, called "*nisipala*", applied after the walls' plastering. The "*nisipala*" thickness varies from area to area, being thicker in the mountain regions (due to more severe winters). Over the dried earth, a thin layer of clay mixed with dung and water is applied, sometimes also including chaff. This operation would be repeated several times per year, especially around the holidays or before family events, when the process described above was repeated. This technique also dates back to the first centuries of our millennia [11].

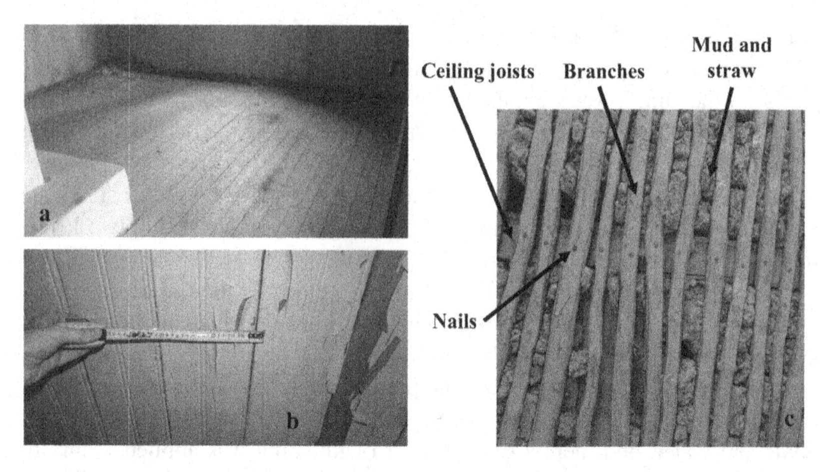

FIGURE 3.18 Floor components of a *paianta* house: (a) ground floor planks sitting on the mudsills; (b) ceiling made of planks; (c) ceiling made of wattle and daub.

In time, this clay floor was replaced with timber plank decking. On the "*nisipala*" a few parallel joists are placed, 3–4 on a room, on which orthogonally, the planks are nails. The planks can be made of fir or, for higher durability, made of oak.

Recently, together with the concrete foundation beams, a concrete slab is cast with the role to support the timber flooring or whatever the owner chooses as a finishing for the ground floor slab.

REFERENCES

[1] Institutul Central de Cercetare Proiectare si Directivare in Constructii, "Studii de arhitectura traditionala in vederea conservarii si valorificarii prin tipizare: Locuinta sateasca din Romania", 1989.

[2] I. Ghinoiu, "*Atlasul Etnografic Roman – Volumul 1 – Habitatul*". Bucharest: Academia Romana Editor, 2003.

[3] R. Group, "Architecture guides on local rural houses". https://oar.archi/buna-practica/ghiduri-de-arhitectura/

[4] C. Miron, "Materiale neconventionale locale pentru energie sustenabila", 2012.

[5] C. Miron, "Final Report regarding potential or existing sustainable building materials that can be or are being produced locally, exiting producers of sustainable building materials or producers of building materials that are interested in expanding their production", 2013.

[6] C. M. G. Călătan, A. Hegyi, C. Dico, "Additives influence on the Earth characteristics used in vernacular construction", in *10th edition of the Int. Conf. Environmental Legislation, Safety Engineering and Disaster Management ELSEDIMA. 39.*, held at Cluj-Napoca, Romania, 18th – 19th of September 2014.

[7] Hegyi, A (Hegyi, Andreea); Dico, C (Dico, Carmen); Szilagyi, H (Szilagyi, Henriette), "Eco-innovative thermal insulation panels, based on organic fibre composite material", *Revista Romana De Materiale-Romanian Journal Of Materials*, vol. 50, no. 2, pp. 205–211, 2020.

[8] A. Dutu, M. Niste, I. Craifaleanu, and M. Gingirof, "Construction techniques and detailing for Romanian Paiant ă Houses: An engineering perspective", *Sustainability (Switzerland)*, vol. 15, no. 2, 1344, 2023.

[9] V. Dragomir, *Conservation and Restoration of Traditional Architecture (in Romanian)*. Craiova, Romania: Editura Universitaria, 2012.

[10] A. Dutu, K. Miyamoto, G. Jole Sechi, and S. Kishiki, "Timber framed masonry houses: Resilient or not?", in *3rdEuropean Conference on Earthquake Engineering and Seismology (3ECEES)*, September, Bucharest, Romania, 2022.

[11] N. Cojocaru, *The Old Timber House in Bucovina*. Bucharest: Meridiane Publishing House, 1983.

[12] A. I. Ciobanel, *The Habitat: Answers to the Questionnaires of the Romanian Ethnographic Atlas: Corpus of Ethnographic Documents/Romanian Academy – Vol. 5: Dobrogea and Muntenia (in Romanian)*, 2nd edition, vol. 5, Bucharest: Editura Etnologica, 2019.

[13] MDRAP, *P100-1/2013. Cod de proiectare seismică – Partea I – Prevederi de proiectare pentru clădiri [Seismic Design Code – Part I: Design Prescriptions for Buildings]*. Monitorul Oficial al României, No. 558 bis/3 September 2013 (in Romanian), 2019.

[14] A. Dutu, H. Sakata, and Y. Yamazaki, "Comparison between different types of connections and their influence on timber frames with masonry infill structures' seismic behavior", 16" World Conference on Earthquake, 16WCEE 2017,paper ID 951, 2017.

[15] A. Duţu, M. Niste, I. Spatarelu, D. I. Dima, and S. Kishiki, "Seismic evaluation of Romanian traditional buildings with timber frame and mud masonry infills by in-plane static cyclic tests", *Engineering Structures*, vol. 167, pp. 655–670, 2018. doi: 10.1016/j.engstruct.2018.02.062

[16] A. Dutu, M. Niste, I. Spatarelu, D. I. Dima, and S. Kishiki, "Seismic evaluation of Romanian traditional buildings with timber frame and mud masonry infills by in-plane static cyclic tests", *Engineering Structures*, 2018. doi: 10.1016/j.engstruct.2018.02.062

Japanese Traditional Timber Frames 4

Japanese traditional residential houses have amazed engineers worldwide for decades. The refinement of the construction details combined with the simplicity of the structural system layout is a unique combination worth understanding and investigating. This chapter refers only to the residential timber houses with a post and beam framework, and infills, which can be still found in Japan and where the carpenters' skills are still transmitted from generation to generation.

Maybe due to the laws during the Tokugawa shogunate, forbidding any display of wealth [1] for residential, except for samurai and nobility classes, the simplicity of the Japanese houses is now a statement of beauty and art.

Be it a teahouse, a mountain refuge, townhouse (*machiya*), farmhouse (*minka*), traditional inn (*ryokan*), or imperial villa, Japanese traditional architecture is mainly based on timber, clay, and straw. Although modern architecture, where concrete has replaced timber, looks very different from traditional architecture, it is also based on the ancient principles of traditional houses.

Hereby, we will focus on the *minka*, the farmhouse type. *Minka* are the traditionally designed houses of the common people of Japan. *Minka* designs are closely related to the climate and character of each region, and the social status of house owners. These conditions have allowed the possibility of a wide range of local types in every region of the country. The design of the house is mainly focused on practical living, and as in the rest of the world, the details have evolved and been adopted after many years of trial-and-error, especially given the fact that Japan experiences earthquakes more often than any other country in the world.

Unlike any other country in the world, the secrets of carpentry are transferred from generation to generation, usually among family members, like sons embracing the same passion as their fathers, but what is interesting is that there are also written guidelines, meaning that carpenters were educated people eager to share their experience and knowledge with others for the sustainability of their traditions.

DOI:10.1201/9781003405375-4

Another reason for that is also the mentality that the carpenters were thankful and respectful to their ancestors, cherishing the knowledge that was passed on to them and doing the same for the following generations.

4.1 TYPES OF HOUSES

Minka or folk houses come in a wide range of styles and sizes, largely because of differing geographic and climatic conditions as well as the lifestyle of the inhabitants. They generally fall into one of four classifications: farmhouses *nōka* (農家), townhouses *machiya* (町屋), fishermen's dwellings *gyoka* (漁家) and mountain dwellings *sanka* (山家). Figure 4.1 shows the *minka* types, according to the infill of their walls.

FIGURE 4.1 Types of vernacular Japanese houses depending on their infill: (a) vertical timber planks infill; (b) wattle and daub infill; (c) cedar bark infill (reproduced with the permission of the Nihon Minkaen Museum).

4.2 CONFORMATION CHARACTERISTICS

Some references present in great detail the Japanese traditional houses' characteristics, but usually they also explain them from the architectural point of view. This book focuses on the structural details and presents them briefly and clearly, so any person with engineering knowledge or not would understand the secrets of the construction details.

The main structural elements of the *minka* frame are presented in Figure 4.2.

The foundations are usually found in three versions: the *ishibatate* method with posts resting on stones, which is maybe the most spread; the earth-fast posts (*hottate-bashira*) in which the lower parts of the posts soon rotted so there are no present examples; and the sill beams used to support the bases of posts (*dodaitate*). Foundation types are shown in Figure 4.3.

As an engineer's eye can easily observe, all types of foundations shown in Figure 4.3 are very flexible. The interesting part is that they could withstand earthquakes reasonably, and the nowadays method with mudsill strongly connected to the concrete foundation actually transfers the energy dissipation into the wall and its connections.

On the foundation, the mudsill, called "*dodai*", is leveled and connected at the corners and wall intersections with standard methods, special joinery being provided for corners, which remain exposed. The posts, called "*hashira*", are

FIGURE 4.2 Structural elements of a *minka* house (reproduced with the permission of the Nihon Minkaen Museum).

Posts resting on stones
Ishibatate (Emukaihouse)

Posts resting on mudsills, which rest on stones
Dodaitate (Hirose house)

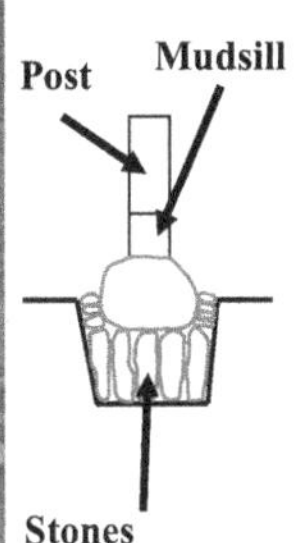

FIGURE 4.3 Examples of foundation types for *minka:* posts resting on stones (left) in Nihon Minkaen; posts resting on mudsills, which rest on stones (center) in Nihon Minkaen and cross-section through a foundation (reproduced with the permission of the Nihon Minkaen Museum).

placed on the mudsills at standard distances and connected to each other by horizontal tie members, called "*nuki*" [2]. The *nuki* penetrates the columns and is fixed with wedges.

In *minka* there may be several types of floors. "*Doza*" is the one in which an insulating layer of bundles of straw is spread over rammed earth, and covered with "*mushiro*" which is a rice-straw mat, or bamboo lattice (like in the ceiling case, Figure 4.7, right) or raised timber flooring with timber planks [3]. Thus, the floors are covered either with earth, wood, or "*tatami*". As stated in [2] the tatami floor represents the living area, the planked area represents the communication and utilitarian areas (hallways), and the earthen floor is a transitional space between inside and outside.

The posts are connected to the mudsills or are sitting directly on the foundation stones, and are connected by mudsills at an upper level (30–40 cm above the stone) as shown in Figure 4.3, left. The top plates called "*ashigatame*" are sitting on the posts, and on them, the slab joists are placed. As stated in [2], if it weren't for the complicated connection details, the frame would not have any lateral stability. An interesting detail is that in our engineering calculations, we often consider timber joints as articulated, but when using a Japanese connection type, this assumption is not valid anymore as some of them are quite moment-resisting and can withstand the horizontal loads, coming from typhoons or earthquakes.

The connections between the posts and the horizontal elements can be quite diverse. Until the roof cover is erected, the operations are quite simple, involving only assembly of vertical and horizontal elements, a simplicity which is often underestimated as being little resistant to horizontal loads. But in fact, the secret stays in the complicated, looking like artwork, connection details. The Japanese carpenters can still make perfect connections with no gaps, unlike, for example, Romanian ones nowadays.

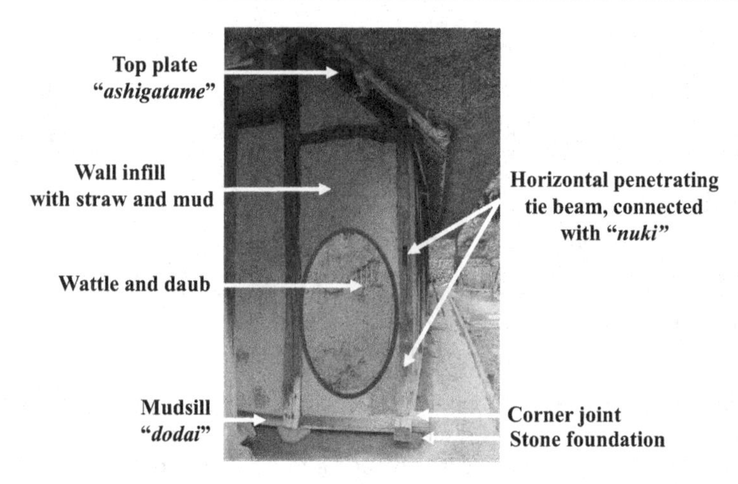

FIGURE 4.4 Mud wall components in the Ito House (reproduced with the permission of the Nihon Minkaen Museum).

Clay mixed with chopped straw is applied on both sides of the wattle and daub infill (Figure 4.4) between the posts, and thus the wall is formed [2]. Finally, the "*tatami*" is laid, which is a specific Japanese floor mat, having the dimensions of 1.8–2.0 m by 0.9–1.0 m. Rooms are most often conceived as a multiple of "*tatami*"s to form a rectangular shape.

Traditional Japanese walls can be made of wattle and daub (with bamboo sticks) (Figure 4.4), timber planks, and thatch (*kaya*). The wattle and daub were the most popular in *minka* due to the availability of the materials, and it is thought to possess excellent fire resistance, thermal insulation, and soundproofing. There are two ways of using wattle and daub infill for the walls, one is when it is less thick than the thickness of the posts, it is left visible and is called "*shinkabe*", and the other is when plaster is applied on the infill and also on the posts, so everything is covered, called "*okabe*". The latter one is used as a fireproofing method.

Engel [2] says that there is no vernacular house without at least one entire solid clay wall. Also, the timber frame is composed of short members, and the connections of the timber elements in a longitudinal direction are frequent, probably due to the ease of transporting it to the construction site.

The roof forms of *minka* houses can be observed to fall into three basic types – gabled (*kirizuma*), half-hipped (*irimoya*), and hipped (*yosemune*). Depending on the layout and dimensions of the rooms, the roof may be a combination of the three above-mentioned types. But another main factor of the roof shape and type is the material used as a cover which may be thatched, tiled, shingled weighted with stones, or cedar barked. In turn, the material used strongly depends on its availability in the houses' region.

The roof framework in *minka* follows the roof form and some examples of layout are shown in Figure 4.5. The framework may have a king post (the central

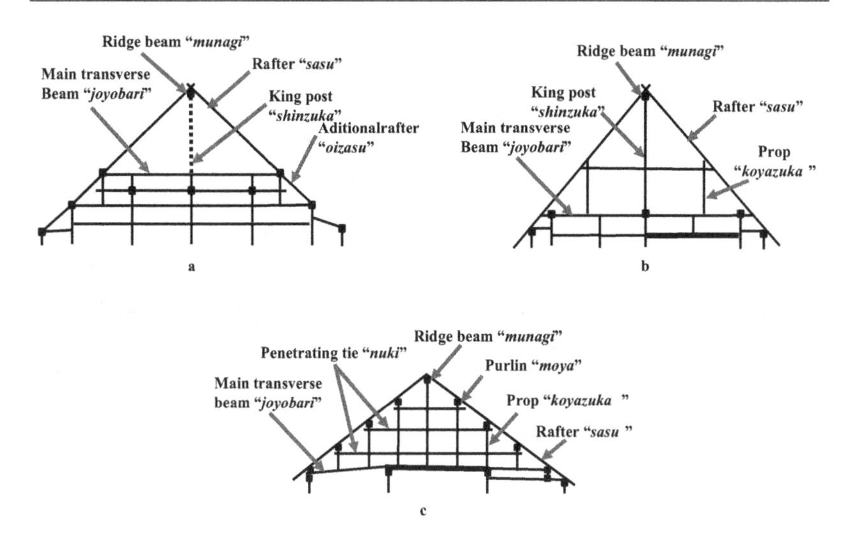

FIGURE 4.5 Roof truss types in *minka*: (a) ridge beam supported by rafters "*sasugumi*"; (b) ridge beam supported by kingpost or post "*odachigumi*"; (c) truss made of props connected with horizontal penetrating ties (Japanese traditional) "*wagoya*".

prop) with different lengths, depending on the height of the roof, and if the roof's angle is higher (i.e. in a colder region, where it snows a lot), short uprights (props) may also be present, resting on the transverse beam, and carrying the roof load to it. The main transverse beam (joist) is usually not straight, but its' shape and defects (crookedness) are used for the benefit of the structure. The cross-section of the main transverse beam is around 200 x 200 mm, but it is not constant along its length. The processing of the timber elements, especially for the roof, implies very few operations such as roughly hewing, and entire trunks are preferred. The elements identified in a real house are shown in Figure 4.6.

On the joists, called "*hari*", the props support the purlins, called "*moya*", and the above rafters, called "*taruki*"[2]. The roof can have a curved shape, due to the low resistance of the rafters, which can bend easily. On the rafters, a layer of planks encloses the roof and represents the support for the clay or latticework, on which the final roof cover sits. The common method to embed the tiles in clay mixed with straw, which is also sitting on bark laid on the rafters, proved in recent earthquakes to be not very effective, as the roof load being so big produces damages in the wall framework which is rather lightweight and usually, due to the tradition, does not present timber braces which can improve the lateral resistance of the wall in earthquakes. Another way to attach the tiles is on a wattle and daub layer, but although it seems better for earthquake behavior, due to the strong winds in Japan (sometimes produced by typhoons) its connection to the tiles is easily broken, and tiles tend to fly away.

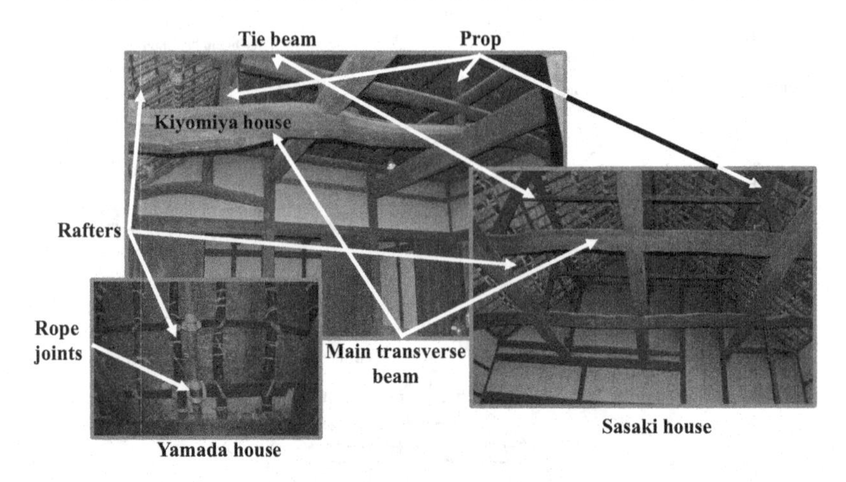

FIGURE 4.6 Roof truss elements in *minka* (Yamada, Kiyomiya and Sasaki houses) (reproduced with the permission of the Nihon Minkaen Museum).

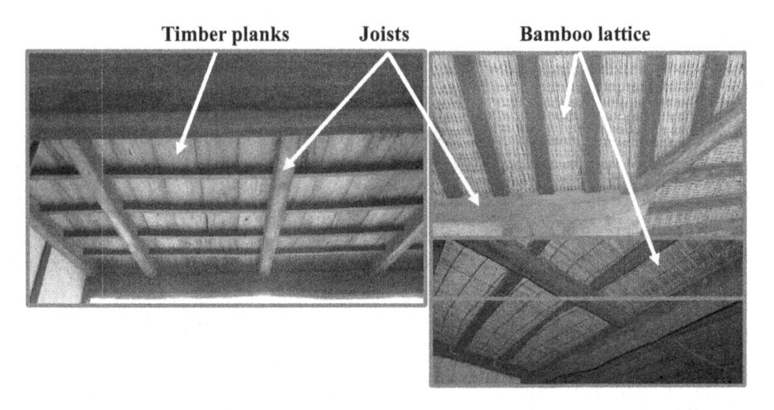

FIGURE 4.7 Load-bearing ceiling with joists and planks (left) in Edo Tokyo Open Air Architectural Museum and with bamboo lattice on beams sitting on joists (right) (upper photo is from Edo Tokyo Open Air Architectural Museum and lower photo is from the Hirose house of Nihon Minkaen Museum) (reproduced with the permission of the Nihon Minkaen Museum).

Just as a measure of the strength of the winds in Japan, I noticed as I was going daily to the Tokyo Institute of Technology Suzukakedai campus, that on the way from the train station to my office building, there was this multistory reinforced concrete building which had small tiles attached on the façade. Although it was well maintained, the wind often broke a few tiles, in the same area usually, and workers repeatedly would replace them.

The ceiling may be classic, with timber planks between the roof joists, with bamboo lattice or suspended (Figure 4.7).

4.3 COMPONENTS AND HOW TO BUILD A JAPANESE TRADITIONAL HOUSE

4.3.1 Materials, Tools and Workmanship

"*Wood*" is the main material used in traditional Japanese architecture, like the bricks and stones are used in other parts of the world. One reason may be the seismicity of Japan, placed in the "Pacific ring of fire" and having earthquakes more often than any other country in the world. It was quickly understood that bricks and stone do not work well under seismic loading. At least not by themselves. My personal experience with the Japanese earthquakes is that on May 28, 2012, when I arrived in Japan to begin my postdoctoral journey in Sakata sensei's laboratory, on the first night spent in my new dormitory room, although greatly jetlagged an earthquake woke me up in the middle of the night. The earthquakes in Japan at that time (a year after the Fukushima disaster) were very often (at least weekly), being aftershocks after the great 9 M_w magnitude earthquake. They felt strange to me, even though I experienced some Romanian small earthquakes, which usually have a big shake initially and then they fade in intensity. The Japanese ones were strong, with big and almost equal sways for at least five seconds. I was shocked that my office colleagues were not impressed at all by these sways that in turn made me paralyzed. But soon I realized that there is nowhere else in the world with higher building safety. The buildings may have swayed but they did not suffer damage from that huge earthquake.

Wood is also abundant in Japan, and in the last decades, the authorities forbid the cutting of its forests and recommended the use of imported timber. Before that, when a carpenter cut a tree, through the cultural Shinto values, he would promise the tree that it would not be wasted, but something useful would result from it, hopefully for future generations to enjoy and cherish. That's how much respect Japanese carpenters have for wood. Even after nine years of moving from Japan back to Romania, I can still remember the smell of the "king" timber type, the *hinoki*, a species of cypress. It is strong (best cut after 200–300 years of growth) and resistant to decay, at the same time easy to work with, and with a regular grain and beautiful color [4]. Other types of local wood are also used, maybe nowadays more, due to the high price of *hinoki*, such as *keyaki* (zelkova), *akamatsu* (red pine) from which the often visible twisted roof joists are usually made, *sugi* (cedar) which is soft, durable, decay resistant and straight-grained wood, and *tsuga* (a yellow spruce) which is straight, flexible and densely-grained [4].

The steps to obtain a wood construction material are:

- Cutting the trees by lumberjacks controlling the cutting machines and falling direction;
- Removing the branches and the bark by trimmers;
- Transport preparation;
- Processing the logs and turning them into squared lumber by sawyers;
- The construction process begins at the carpenters.

The houses' construction sites were led by a head carpenter, who had the role of manager too. This position involves many skills; besides carpentry, he is actually designing the house using basic equations and rules of thumb [4], and also supervising the entire construction process of the house. The other carpenters involved were called "*daiku*". The carpenters in Japan have a code of conduct, which teaches them as young as nine years old, that they must do their job with respect and without wasting natural resources, as skillfully as they can.

A great master carpenter (Tsunekazu Nishioka, born into a family of five or six generations of carpenters) advised that the selected timber should be used preferably in the close location of its cutting and in the same orientation it has grown in originally (N-S-W-E), as it grew in harmony with exterior forces, and they should be similar when it is used in a building to be able to make the best use of the tree's properties. Master Nishioka also said that if a tree during its lifetime is subjected to strong winds from the east, it will be prone to twist clockwise [4]. Another tree may bend because it has released tension caused by heavier branches on one side as a result of how the sunlight reached it [4]. If such wood is combined with wood having opposite behavior, a strong structure may be created in the end and thus the elements, not perfect by themselves, help each other in the whole picture. Woodworking includes a lot of wisdom. Moreover, the best tree for construction is the one growing on the top of the mountain. Trees used in this way survived sometimes more than 1300 years.

Nishioka san was saying about the situation in Japan that nowadays, academia complicated a lot the things that were once simple, and in the last 100 years it overlooked and thus lost 13 centuries of accumulated practical knowledge. This situation is not only happening in Japan but probably worldwide. Now, many scholars are researching clues for the secrets of such great workmanship of the woodwork, but there are facts like the personal hand of the maker that cannot be ever reproduced or described in words. In old times, the carpenters communicated a lot between them and were competing, thus being driven to innovate and be as best as they could be. Nowadays prefabricated wood construction prevails ahead of traditional, thus manual, ways, and young people are advised and are eager to enter the "salary-man" world, which mainly involves office work.

The master carpenter has a central role within this wood crafting. A very important job for him is to train apprentices, to maintain the continuity of carpentry skills. But what is very interesting is that things are not fully revealed to the apprentice, but mainly allowed to explore his own instincts and draw his own conclusions [4]. This is exactly what I experienced in Sakata Laboratory, under the supervision of Professor Hiroyasu Sakata. I was allowed to study, think, make mistakes, repair them, and learn as much as I have to learn. And all this time, whenever I had problems I couldn't overcome, Sakata sensei was there to help, unblock the way I wanted to go, but only after I had done everything possible to unblock it myself. The experience of work done by himself is maybe the most precious lesson for a carpenter, as he knows what and how he had done the work, sees it in time, and is able (if it is within his purpose to become better) to identify and criticize himself, understand what he had done wrong and do it better next time.

The master carpenter must be compassionate, be a true leader, and possess communication skills. Respect and humility towards the wood material and the work on it are also the key elements in becoming a great carpenter.

The first jobs for a young apprentice involve simple things like helping to lift heavy members, sweeping the floor, and getting something to drink for the master and colleagues, all these aiming at fitting into the team and practicing concentration and exactitude [4]. Then he is gradually learning to use the tools and assisting senior carpenters with their tasks. It takes about seven years to become capable of shaping a complex piece when reading the drawings and templates, and about 15 years to make the templates himself. A senior carpenter can distinguish the region where a tree comes from after its wood's smell. Once you understand "sun, wind and rain", a carpenter once said, you will understand wood [4].

Japanese carpentry manuals appeared in the Edo period (beginning of the 17th century), but many master carpenters kept their own manuals which survived over time.

Some rules of thumb passed on by master Nishioka in [4] will be briefly presented further, based on observations throughout the centuries on ancient building's behavior in time.

The ideal building should have a mountain to the north, a river to the east, a lake to the south, and a straight road to the west. The best argument for it is an actual temple surviving for many centuries in the exact same conditions. If the river is missing, seven willow trees (a river-loving tree) should be planted instead. Instead of a pond, seven paulownia trees can replace it [4]. Instead of a road, seven plum trees, and instead of a mountain, seven *enju* trees.

The orientation of a cut timber is to be kept within the new building. Trees that receive strong sunlight due to their location on a southern slope are best used as columns on the southern face of the building. Since a tree's natural property is to resist gravity loads, it is thicker at the base and thinner at the top. Thus, it is better to keep the vertical relationship and base end of the tree to be used down and the top always up. When possible, horizontal or inclined elements should be joined top-to-top.

The location of the tree gives us the strongest ones above the midpoint of the mountain, where they receive the best sunlight and the greatest air circulation. The trunks are thicker since the trees can branch faster and fully and develop the necessary girth to support a large number of leaves or needles without being cramped by the neighboring trees. These trees should be used for columns. But the trees below the midpoint of the mountain are forced to compete for sunlight, thus growing higher before branching. These are usually thinner and without knots and should be used for exposed members which require a fine surface. The valley trees produce lower quality timber, having a high moisture content, and are recommended to be used for ceiling boards, or other parts which do not require strength or fine aspect.

Living of the wood even after it is cut comes with movement and shrinkage, as a consequence of its exposure to temperature and moisture. The biggest deformation occurs perpendicular to the grain when viewed from the end, that is, radially. When processed, the timber should be cut in such a manner to minimize the shrinkage effects, thus the changes can be predictable and taken into account in the design. Elements with twisting grains should be used in opposing pairs, so that the movements counteract each other.

The tools are crucial for the result in a woodworking process. This is also a detail in Japan, known for its great metallurgic industry, creating machines and auxiliary pieces for any other industry. It started with the carpentry tools, made carefully to be able to craft any shape necessary for the houses.

Most of the initial preparation of the element is done by basic tools like axes and adzes.

The first step is to examine the timber, observe the wood grain and knots positions, and try to predict how the element will deform in time. The carpenter must understand how the final element will look like and try to reach that objective while taking into account the timber's features. This step is maybe the most important for a carpenter, as it requires many sophisticated skills, such as imagination, visualization, or intuition [4].

Then, primary milling is done in a nearby sawmill, and after that, the lumber is transported back to the workshop, where smaller pieces are roughly cut using large-capacity band saws and table saws.

Template timber is used to start processing the timber logs in the next step, in a finer cut. In this step, steel squares are used, inkpot and line, and a bamboo marking pen. The ink cannot be erased, and this has helped people over the centuries to understand the thinking of the carpenter who made them, when they want to restore the work.

The power tools are not avoided in the Japanese carpentry, but hand tools are encouraged to be used by a young apprentice, to understand the way timber resists carving, to understand its properties, and to practice his skills. When an electric blade does the work for him, it is hard to understand the stiffness and flexibility balance of the timber material. In fact, everything can be done with a chisel, hammer, saw, and bubble plane. The apprentice's success is also based on his confidence in his master (sensei) and colleagues, and also the entire system of carpentry. Cutting, trimming, and smoothing of an element is done in this order several times.

The common hand tools involved in the carpentry process are divided into several categories such as: marking tools, axes and adzes, saws, chisels and slicks, planes, spear planes, hammers, and sharpening stones. A very interesting difference is in the adzes' shape, they have the same function as the Western ones but do not look the same (Figure 4.8). The Japanese carpenters prefer handmade steel parts, recycled from old metal tools, believing that the ore collected in the old times has a better quality than the ore available nowadays. The spear plane, although theoretically doing the same job as an electric plane, gives the timber a smoother surface, more water resistant in time [4].

FIGURE 4.8 Adzes in Japan and in Romania (reproduced with the permission of the Nihon Minkaen Museum).

4.3.2 Foundations

The traditional residential buildings usually rest on foundation stones placed on well-compacted soil, and then compacted again after the actual erection to ensure stability. This process is called "*jigvo*" and it was, in fact, a teamwork in which relatives, local people, and skilled workers were involved. The owners organized the *jigyo* as a celebration. This process was popular in Japan until the mid-1950s.

Initially, based on the set dimensions, wooden stakes are driven into the ground at even intervals around the perimeter of the site for the new building and linked by penetrating ties (*nuki*), which are usually nailed in position and leveled. The span between the posts used for the building is marked on the *nuki* in ink. From the marked points strings are stretched across the site in both orthogonal directions creating a string grid. The nodes of the grid, where the strings intersect, indicate the positions of posts, and consequently where foundation stones should be placed [3].

Then the holes in which the stone foundations are sitting, are dug in the ground, the process being called "*negiri*". The first layer in the holes is made of smaller stones, with a diameter between 15 and 25 cm, called *wariguri ishi,* serving as a capillary water barrier.

For big houses, a pestle was used for compaction of the soil which was so big that it needed handling by eight people. Sometimes scaffolding was erected to support it and ropes were used to lift and lower it as needed. Some skills were necessary to do this job, so it was mostly done by workers, but villagers could also get involved to help. When the soil was considered compacted enough, by an experienced head carpenter, fine gravel was inserted into the holes, and it was also compacted. The foundation stones were placed on the gravel, and the area surrounding it was half buried with sand. Finally, more compaction was done.

Nowadays, foundations are made of reinforced concrete, for earthquake safety reasons, and on them, the stones supporting the columns are installed. The stones are sometimes concave in shape so the water can drain easily from them, avoiding the rotting of the bottom part of the timber columns [4].

4.3.3 Structural Framework (Walls)

This next part is done by carpenters. Initially, they would carve the posts, beams, and ties and then assemble them on-site. In rural areas, the villagers themselves, led by the head carpenter, would work and do that, but in towns, usually only carpenters would do the work.

FIGURE 4.9 Japanese mortise-and-tenon connections with dowels: between tie beams and post and between post and mudsills.

The framework for the walls started with the erection of the posts, connecting their bottom parts (the carved tenons) with mortised floor beams and placing also the penetrating ties (*ashigatame nuki*). At the upper part, they were connected using other penetrating ties at lintels and ceiling height. The timber elements were mainly connected by solely timber joints, and rarely nails were also used. The main lower frame of the building, fitted together in this way, is referred to as the *jikubu* or *jikugumi* [3].

There are many types of timber connections in Japan. Some of them are shown in Figure 4.9, joining the post with the tie beams and with the mudsills, by mortise-and-tenon with timber dowel.

4.3.4 Roof Framework

After the walls' framework was erected, the roof framework (*koyagumi*) was assembled (Figure 4.10). This frame supports the roof loads and if it was thatched, *sasu* (inclined roof supports) or rafters were mainly used. When the roof was made of shingles or tiles, the *wagoya* or Japanese truss was more often used (Figure 4.5c).

The roof frame was considered completed when the ridge beam was placed. The ridge-raising ceremony (*mine-age iwai, jotoshiki*) is a tradition in Japan, held to thank all those involved for their efforts and to pray for a successful

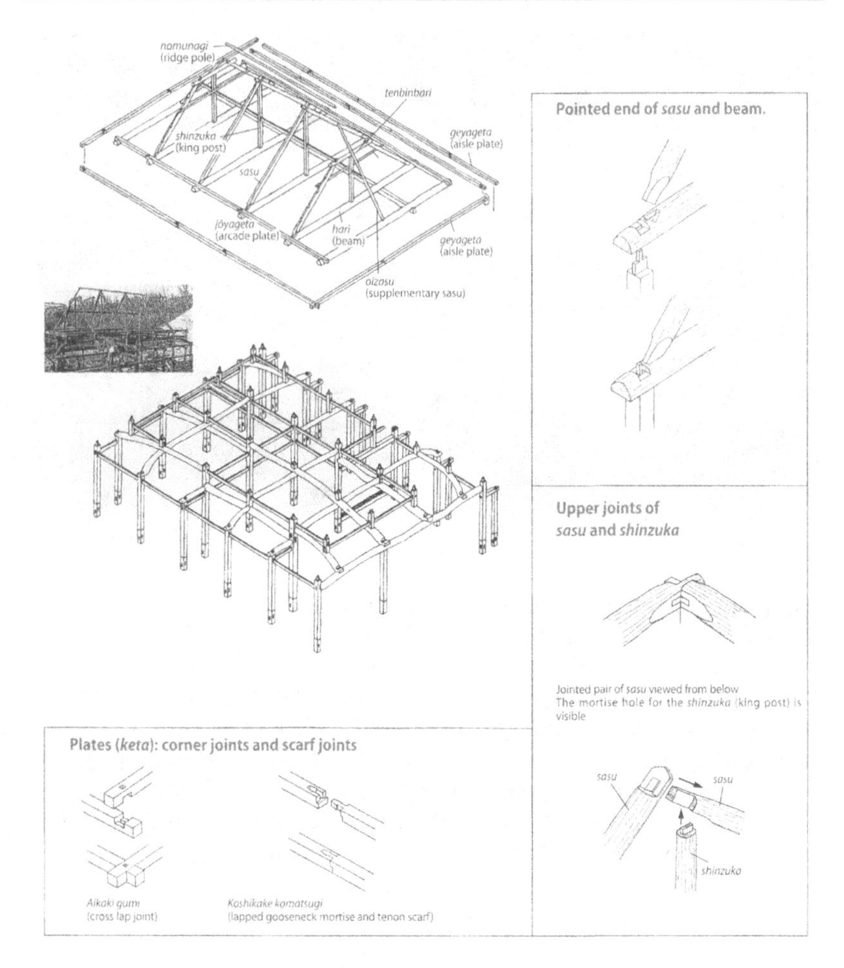

FIGURE 4.10 Making the roof framework, *koyagumi* (up) and the main lower frame, *jikubu* (down) (reproduced with the permission of the Nihon Minkaen Museum).

conclusion of the work and the family's future prosperity [3]. After this, the carpenter passed responsibility to the thatchers or tilers, and the plasterers to finish the walls' infills. When their work was complete, the carpenter returned to make the internal finishings.

4.3.5 Roof Covers

Most of the traditional houses in Japan, especially the farmhouses, have thatched roofs of *kaya*. As explained by [3], the term *kaya* includes all kinds of straw that can be used for thatching roofs such as Miscanthus (*susuki*), reed, *orugaya*, *megarugaya*, wheat straw, rice straw, and others. The advantages of thatched roofs were that all the necessary materials – *kaya*, bamboo, and straw rope, etc. – were available locally and that the execution could be easily done by the villagers themselves, with no special skilled workmanship. As a consequence, thatched roofs in each area show significant regional characteristics in terms of the selection of materials and design of roof ridges and eaves, which are very interesting features of traditional houses worldwide.

The steps of thatching a roof are further presented as described in [3] and shown in Figure 4.11, then the final result is presented in Figure 4.12:

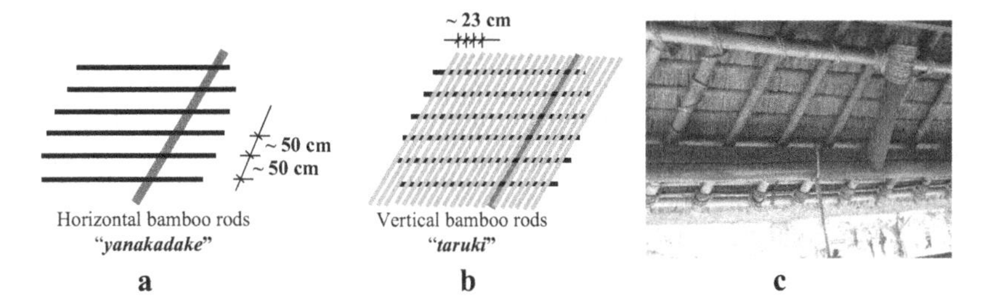

FIGURE 4.11 Creating the support for the straw bundles application (photo (c) is from the Edo Tokyo Open Air Architectural Museum).

FIGURE 4.12 A thatched roof traditional house with flowers on the ridge of the Kiyomiya House in Nihon Minkaen Museum (reproduced with the permission of the Nihon Minkaen Museum).

- Horizontal bamboo rods called *yanakadake* are attached to the rafters (*sasu* and *oizasu)* at intervals of about 50 cm. Then vertical bamboo rods (*taruki*) are laid over the horizontal ones (*yanakadake)* at about 23 cm intervals. The connections between the horizontal and vertical bamboo rods are made with straw rope.
- On the vertical bamboo rods *(taruki*) are placed horizontally smaller diameter bamboo rods *(komaidake)*, either full section or split, at intervals of about 15 cm. On the contour of the eaves of the roof, long thicker bamboo rods are horizontally laid.
- High-quality straws *(kaya)* are chosen to finish the underside of the eaves because they remain visible.
- The straw bundle (*kaya)* to be used for the main part of the roof is prepared by removing the leaves and making their length even.
- The prepared straws are placed vertically, and on top of them other bamboo rods called *oshihokodake* are placed horizontally, and several layers are made in the same way. The edge of the roof is now complete.
- The straws are laid from the eaves towards the ridge beam.
- The bamboo rods on top of the layers of straw bundles (*kaya*) are connected to the wattle and daub support by straw rope by means of large needles (in Japan made of bamboo, while in Romania of steel) and two men, one on the roof and the other inside the roof space, push back and forth through the bundles of thatch until the thatch is tied in position.
- After several layers of straw bundles are placed, their ends are cut and compacted with a tool called *ganki*, to obtain a smooth and even surface.
- The straw is bent over the ridge beam. Several layers of straw form a base, which is then wrapped in cedar bark and more bundles of straw. In some houses, green grass or flower seeds are planted in this area.
- In the end, the straw is trimmed starting from the ridge beam towards the eaves, to obtain a smooth roof cover.

4.3.6 Infills and Finishing on the Walls

The wattle and daub infill is done by knitting bamboo within a post and beam already erected frame and then applying several layers of clay on it.

The base layer bamboo is not split to allow the choice of making either *shinkabe* or *okabe*, so the full round section of the bamboo is used, usually with 20–30 mm diameter. Here are the steps for executing the wattle and daub infilled Japanese style wall, also shown in Figure 4.13:

- The posts and beams (top plate and mud sill if it exists) are carpentered, posts are erected at 910 mm interaxle, connected through mortise-and-tenon joints usually, but other types are also possible;

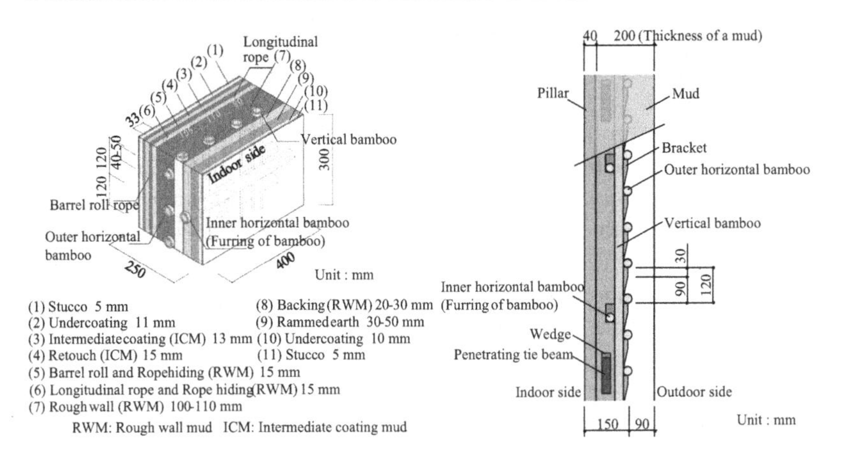

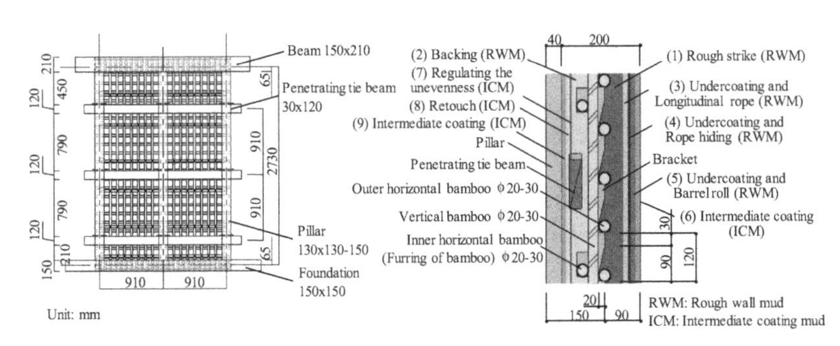

FIGURE 4.13 Details of an inner mud wall and Cross-section of an inner mud wall around a post [5] (reproduced with the permission of Prof. Yokouchi and licensed under Creative Commons Attribution-NoDerivs (CC BY-ND)).

- The tie beams are erected penetrating the posts and with a maximum distance between them on the vertical direction of 790 mm, 910 mm interaxle;
- The inner horizontal bamboo is hung between the posts on both sides in a frame;
- On the outside, vertical bamboo is installed;
- The outer horizontal bamboo is placed into a sawblade-shaped bracket cut out of the post so that the weight of the mud is transferred from the outer horizontal bamboo to the posts;
- The intersections of bamboo are tightly tied with straw rope to produce a solid substrate [5];
- At almost the same time as the bamboo knitting, the clay is collected, and it is sifted through a sieve and left to rest for one week;
- The clay is kneaded, with water and sand;

- A rough but thick layer of clay is applied on the wattle and daub infill made of bamboo knitting; thus the thickness of the wall reaches 100–110 mm;
- Depending on the side of the wall, several other layers of clay are applied with at least four weeks of waiting time between them;
- For the exterior side, the layers are applied in the following order: rope hiding with a thickness of 15 mm; barrel roll and rope hiding 15 mm; a layer of straightening the wall of 15 mm, another layer of clay 13 mm; finishing support (also with clay) of 11 mm and finally 5 mm the plaster;
- For the interior side of the wall, the following layers are applied: a support layer of 20–30 mm, rammed earth 30–50 mm, finishing support (also with clay) of 11 mm and finally 5 mm the plaster;
- The resulting thickness of a mud wall is about 200 mm, and with the post, 240 mm.

In constructing parts of a wall where the cross section becomes smaller owing to the frame, the rough wall, longitudinal rope, barrel roll, and straw rope are densely arranged, to maintain the integrity of the mud. Furthermore, when increasing the wall thickness, a rough wall mud is used, with intermediate coating mud thinly plastered to reinforce the fixing of the rough wall mud and to smooth the wall surface. Subsequently, retouching is performed, and the finishing materials are plastered.

Two types of mud – rough wall mud and intermediate coating mud – can be found. The rough wall mud is clay mixed with straw that is approximately 50 mm long and kneaded with water. The intermediate coating mud is clay mixed with sand and fibrous straw that is approximately 20–30 mm long and kneaded with water.

For the infill with the timber planks, there are no grooves on posts or beams and the planks are nailed directly to the penetrating wall ties ("nuki"). Planks are inserted directly against each other, and vertical battens called *me ita*, fixed on the inside of the wall, are used to seal the gaps between the boards [3].

The main steps of plastering walls are:

- preparing the base frame *(shitaji tsukuri)*,
- rough wall (*arakabe*),
- middle coat (*nakanuri*),
- final coat (*uwanuri*).

The middle and final coats are more specific to temple buildings and *minka* is usually left with only the rough wall, without other layers of plaster.

REFERENCES

[1] N. Murata, and A. Black, *The Japanese House. Architecture and Interiors*. North Clarendon: Tuttle Publishing, 2000.

[2] H. Engel, *Measure and Construction of the Japanese House*, Kindle edition. Tuttle Publishing, 2020.

[3] N. Minkaen, *Nihon Minkaen – Japan Open- Air-Folk House Museum*. Kanagawa Prefecture, Kawasaki city: Nihon Process, 2011.

[4] A. Brown, *The Genius of Japanese Carpentry. Secrets of an Ancient Woodworking Craft*, Revised. Tuttle Publishing, 2013.

[5] H. Yokouchi, and Y. Ohashi, "Repairing and recovering structural performance of Earthen walls used in Japanese dozo-style structures after seismic damage", *Journal of Disaster Research*, vol. 13, no. 7, pp. 1333–1344, 2018. doi: 10.20965/jdr.2018.p1333

Seismic Behavior of Traditional Timber Frames

5

5.1 EARTHQUAKE PROOFS

In recent years, many earthquake reports indicated the good deformation capacity and resilience of timber frames with infills, such as in Europe for:

- The Greek traditional timber frames with infills called "*Xylopikti Toichopoiia*" in the Lefkada earthquake in 2003 [1] and the Samos earthquake in 2020 [2];
- The Turkish *himis* houses in the Kocaeli earthquake in 1999 [3, 4] and the Samos earthquake in 2020 [2];
- The *paianta* houses in the Vrancea earthquake in 1977 [5, 6];
- The Italian *casa baraccata* in the Messina earthquake 1908 [7].

Then, in Asia for:

- The *ChuanDou* houses in the Lushan earthquake in 2013 [8];
- The Nepalese traditional houses made with stones, timber posts and seismic bands in the Ghorka earthquake in 2015 [9];
- The *centibera* buildings in the Sikkim earthquake in 2011 [10];
- The *dhajji dewari* in the Kashmir earthquake in 2005 [11];
- The Miyanmar traditional timber frames with brick masonry infills in the Chauk earthquake in 2016 [12].

In the Americas:

- The *Gingerbread* houses in Port-au-Prince in 2011 [13] and *kay peyi* houses in the Petit-Trou-de-Nippes earthquake in 2021 [14];

DOI:10.1201/9781003405375-5

- The *quincha* houses in the Pisco earthquake in 2007 [15];
- The *bahareque* houses in the Muisne earthquake in 2016 [16].

Many researchers recently started to observe and study the surprising behavior of such houses, not because they went through the seismic events without damages, but because of their capacity to protect the life of the inhabitants and the repairs necessary to make the houses inhabitable immediately after the earthquake being easy, often using the same materials and rebuilding the infills as they were without the financial intervention of the government of international aid.

And this is the reason why this book appeared, because nowadays engineers do not know much about traditional vernacular houses, while they sometimes prove to be safer than modern constructed homes, executed poorly, such as reinforced concrete frames with masonry infills.

Analyzing the published graphs with the spectral acceleration *SA* (in *g*), function of the period *T* (in sec), for all the earthquakes mentioned above, it is interesting to see that except the Muisne earthquake in which the maximum recorded value of the peak ground acceleration was 3.5 g [17] corresponding to a period of 0.1 sec, for the other earthquakes peak ground acceleration (PGA) didn't exceed 0.7 g, for a period of 0.14 sec (as calculated and obtained by dynamic ambient tests [18] for a traditional *paianta* house). This means that such buildings are not subjected to a "proper" frequency content earthquake, thus not being too much affected due to their low period value, usually being lower than the dangerous values recorded for an earthquake. Buildings with higher period values are usually more affected, but there are proven exceptions like the Ecuador, Muisne earthquake, which showed that timber frames with masonry infills behave reasonably, even though they were not very well maintained previous to the earthquake.

5.1.1 Romania

For the last strong earthquake in Romania, in 1977, although there are no available photos with evidence, within the national reports [5] a small statement exists related to traditional buildings (with one floor and rarely two stories) made of local cheap materials, specific to different geographical areas (in general made of stone and timber in mountain areas, and *paianta*, adobe and earth in the plain areas). The configuration of such buildings, made intuitively – due to tradition – by local builders (most of them have rectangular, regular shapes in plan), was according to the seismicity of the area, thus avoiding significant damages and casualties during strong earthquakes. Among different traditional structural types, the best behavior was observed at timber and *paianta* houses, and the worst at the adobe and earthen buildings, for which heavy damages were reported and even collapses. A confirmation of the good behavior in the March 4 earthquake in 1977 of the traditional rural buildings was offered by the more than 100 rural houses from the National Village Museum in Bucharest (the city in Romania hardest hit by the earthquake).

Another report [6] presents that among the rural dwellings, identified in six typologies, the main damage types were:

1. Timber structures didn't suffer any damages or degradations;
2. Stone masonry structures had minor cracks, depending on the quality and age of the used mortar;
3. Brick masonry structures showed collapses, and severe damages (cracks, dislocations, expulsion of untied walls) depending on various factors;
4. Adobe brick structures suffered collapses, significant damages, cracks and dislocations of cornices, gable walls, and columns;
5. *Paianta* structures (mixed timber and clay) had good behavior, except for cracks in lintels and infills;
6. Rammed earth and earthen structures had various behaviors, some of them showing severe damages (enough to make them unfunctional, but not producing casualties), and others not showing any damage (in the same location).

5.1.2 Japan

The recent earthquake in Kumamoto, in 2016, which was, in fact, a series of two earthquakes, one after the other, in a 28-hour interval, showed the vulnerability of traditional timber houses, especially due to the heavy ceramic tiles roofs. The damage investigation report shows many houses with the first story crushed by the upper one, thus a clear soft-story mechanism collapse. Some houses have traditional mortise-and-tenon connections which failed easily, while others have modern hold-down connections, and their capacity was also overcome by the seismic forces.

Figure 5.1 shows a traditional *minka* house with mud infill walls made of wattle-and-daub with bamboo sticks. Images of the infills deformed or collapsed in the out-of-plane direction demonstrate the little influence they have in the lateral resistance of the house.

In the 2011 Tohoku Region Pacific Offshore Earthquake, the traditional townscapes and *dozo*-style structures of the Kanto region were also seriously damaged.

FIGURE 5.1 Damages to the traditional timber houses after the Kumamoto earthquake, 2016 (reproduced with the permission of Prof. Akira WADA).

An investigation was conducted by Professor Matsumoto from Tohoku University [19] on both modern and traditional timber houses after the 2011 earthquake in Fukushima, Miyagi and Iwate prefectures. Among the 16 buildings, there were no buildings that collapsed. Damage observed varied from moderate to heavy level and included broken or shifted structural elements (including beams, lintels, etc.), corner connection damages, sliding of the columns on the stones/concrete foundations and damage to earthen walls and plaster walls. Most of the damage consisted of cracks in the infilled walls, and the heavy ones involved out-of-plane falling of the infills.

Another type of timber frame can be found in Japan, resembling the western half-timber houses or the French *columbage*, and having as infill burned clay brick masonry. During the transition from the Edo era to the Meiji era, French engineers like Edmond Auguste Bastien embarked on the construction of the Yokosuka Naval Shipyard in 1865 and the Tomioka Silk Mill in 1871 [20]. With the onset of international trade, accommodations for foreign nationals, such as the Jugobankan Building in the Former Kobe Foreign Concession (1881), were also erected. Simultaneously, Marc Marie de Rotz, a Christian missionary, oversaw the design of their religious facilities, including the Former Latin Seminario (1875). Consequently, timber-framed masonry found application in relatively large public edifices during this time. The primary architectural characteristic during this period involved brick masonry walls encasing the interior timber frame. But other combinations of timber and masonry were also built, such as the masonry being confined by the timber or the masonry being on the exterior of the timber frame.

In 1891, the devastating Mino-Owari Earthquake (Nobi Earthquake) inflicted significant damage on masonry structures. As a result, the use of timber frames with brick masonry infills (TFM) became less frequent in public buildings. Nevertheless, with the increase in domestic brick production, TFM saw a resurgence in private residences and corporate structures, exemplified by the Former Kabuto Beer Factory (1898). Some TFM warehouses integrated traditional Japanese construction methods into their design. The central feature of construction during this era was that the brick walls served as an outer layer for the timber frame.

In 1923, the Tokyo-Yokohama earthquake (Kanto Earthquake) caused widespread destruction of masonry buildings. The subsequent year witnessed the implementation of stricter regulations under the Urban Building Law of 1919, making it more challenging to construct masonry buildings in Japan. This led to a rapid decline in the use of TFM. These stringent standards were subsequently incorporated into the Building Standard Law enacted in 1950.

As a result, in 2012 when I joined the Sakata Laboratory in Tokyo Tech, companies executing brick masonry were so scarce, and mainly building fences, that I had to build the first masonry specimens by myself.

5.2 EXPERIMENTAL PROOFS

5.2.1 Romanian *Paianta*

When asked during the field investigations about their houses some owners recall that their traditional house withstood some damages in the last strong seismic event (1977), within the whole range from small (plaster repair) to extensive reparations (infill replacement, to sometimes demolition and reconstruction). Other owners say that in the 1940 earthquake in Vrancea, their unreinforced masonry houses collapsed, and after seeing their neighbors' *paiantas* surviving the event, they decided to reconstruct their houses with the typology as their neighbors.

The seismic behavior of *paianta* houses was also investigated experimentally recently and the results will be briefly presented further, while detailed results are published in [21, 22]. For the first time in Romania, when I came back from Japan, experimental tests led by me were performed on wall specimens of *paianta* houses and were executed having as a reference an existing traditional house. Starting from 2015, two research projects were dedicated first to the evaluation of the existing *paianta* structures (TFMRO project) and then to their strengthening (STRONGPa project). But many variations of the construction details are possible, as mentioned before, this being a feature of traditional houses all around the world. However, on the selected configurations, the results show that the infill offers the main timber frame a greater stiffness and also increased maximum capacity, no matter the infill type, when resisting lateral loads, than in the case without the infill. The infill also restricts the out-of-plane behavior of the braces, which in the case of the walls without infill showed a significant slippage at the connection with the frame, although it is quite sensitive to out-of-plane direction itself.

Sometimes the infilled walls were moving in the out-of-plane direction even before the experiment, during the setup. So there usually is no connection between the infill and the timber frame. During the experiments on walls, which were static cyclic, at large shear angles (>0.05 rad), the infill panels for the fired earthen bricks showed a slight movement perpendicular to the testing plane. When testing infilled walls with and without braces (although different types: 2 versus 3), it was interesting to see that the presence of the braces doubles the maximum force when subjected to the same lateral and vertical load combination.

The position of the timber bracing's connection to the post or directly into the upper post-beam joint is also an important detail. When connected to the upper beam and post (upper corner in the wall) the brace gives more stiffness than the layout with lower bracing (~ 300 mm below the corner joint), but the maximum capacity is at similar values (Figure 5.2)! It is not clear why in most of the existing houses the bracing is lower (Figure 5.3), as it may be that the bracing was just the temporary support of the column and was not considered structural, but it also may be for economic reasons (shorter elements = cheaper and easier to get). It is amazing that all of the *paianta* houses have the brace in the lower position.

According to the current building code [23], timber structures are perceived as low dissipative structures, and their connections are generally considered pinned. But for the *paianta* structure, the infill is a dissipative zone, and based on experimental testing on connections (even without the metallic fasteners), joints proved to be semi-rigid, and not pinned. To support this, a bare timber frame was tested in Japan and all the connections were cross-halved, and it supported an 18 kN maximum lateral force, although based on the engineering manuals, considering its connections pinned, it was supposed to have no resistance [24].

Within the TFMRO and STRONGPa projects [25], for which I was the project director, several wall configurations were tested, consisting of the types

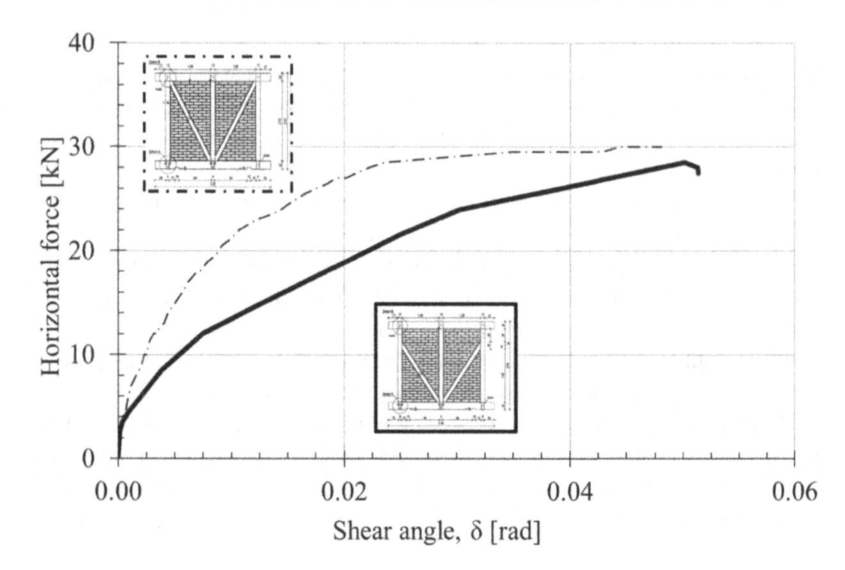

FIGURE 5.2 The influence of the position of the diagonal in the wall's envelope curve (type 1 *paianta*).

1, 2, 3, 4, and 5 (please check Chapter 3 for the typologies description). Type 1 had a fired mud brick infill, and besides testing several specimens to check the variation of the results, the position of the diagonal bracing was investigated too. Type 2, having a timber frame (posts and beams) identical to type 1 but with a wattle-and-daub infill (Figure 5.3c), was tested with and without braces (low diagonal). Several type 3 specimens were tested having a timber frame (posts, beams and low connected bracings) exactly as the type 1 but infilled with mud and straw and on both sides of the wall horizontal timber strips worked as a formwork for the mud infill. Figure 5.3 shows the specimens that were tested in-plane in a static cyclic regime. The jack that was used to apply the horizontal load had a maximum stroke of 220 mm, and it was difficult to observe the real behavior, at least in the plastic range, and the collapse could not be attained.

In STRONGPa project 5 x 150 mm nails (plain) were used, while in TFMRO screw nails were used, two pieces for each connection. Although theoretically the screw nails should give stronger connections, when observing the bending test results on connections, it seems that the ones with nails (three identical ones) are stiffer and stronger than the ones with screws. However, there are other parameters to take into account, such as the mud mortar strength. The type 1 tested within the TFMRO project was tested one month after construction. But the type 1s of the STRONGPa project were tested after two months. The mortar usually becomes stronger with time. In STRONGPa, after testing the walls in the central position of the reaction frame (having a maximum stroke of 100 mm to the left and 120 mm to the right), the specimen was shifted to the maximum left and tested for the last half of the cycle in the positive direction until 220 mm top displacement. This allowed us to observe the damages in the final stage of the wall, usually bending failure of the left post and all the infills' cracking. For the type 1 specimens, the out-of-plane behavior of the infills was significant.

Figure 5.3 presents the envelopes obtained after the static cyclic tests on walls, and for walls photos (a) to (e), the envelope curves on the positive loading is shown to get a quick image into the difference in capacity among the tested specimens. One can see that the type 5 has the best behavior in terms of initial stiffness and maximum force. For type 1 with the lower brace, its

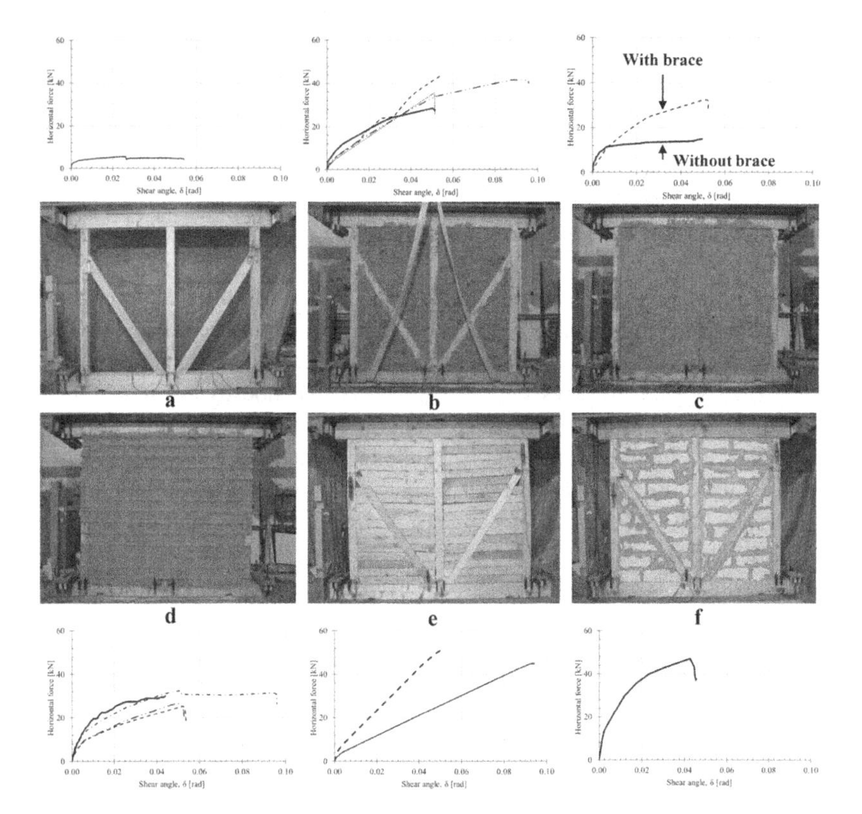

FIGURE 5.3 Wall specimens tested in TFMRO project and STRONGPa project: (a) type 1 *paianta* with low diagonal and no infill; (b) type 1 *paianta* with low diagonal and infill; (c) type 2 *paianta* with and without diagonals and infilled; (d) type 3 *paianta* with full (thick line) and low (the dashed lines) diagonal and infill; (e) type 4 *paianta* with low diagonal and infilled; (f) type 5 *paianta* with low diagonal and infilled.

position clearly reduces the stiffness until large displacement when the bracing is activated, working in parallel with the masonry struts and finally the specimen reaches a similar value of the maximum force as the type 1 with full brace. The infills' strength in compression, especially above the connection between the braces and the post, where the brace tends to slide, is the defining element in the total capacity of the wall. If the infill is strong in compression and blocks the sliding of the brace, then the stiffness and capacity of the wall is higher. For soft infills like in the case of type 3, the stiffness and maximum force are also reduced.

An interesting result was for the wattle-and-daub infilled specimen (type 2), which the legend says would be the strongest one, due to the interlacing connections to the frame, which it was said (by regular people, even from old times) that has a confining effect on the frame. In fact, the experiment showed that it has half of the maximum strength obtained for the other specimens, when it has no bracings. The traditional version of this house appears in some documents to contain no bracings [26], but sometimes in the existing houses there are bracings, either lower braces or even small braces, at the bottom of the posts, connecting into the post at 1/3 of the post's height. When it has braces, the capacity is slightly lower than the one of type 1.

The mud brick infill of type 1 showed in all four wall tests that it cracked even before the beginning of the experiments, this being according to a real situation when clay mud cracks after application. That is the reason for which every year around holidays and family events, the houses are repaired by applying a small layer of new mud to fill in the cracks.

For type 3 (mud and straw infilled and horizontal timber strips), not many damages were observed, except cracks in the mud infill especially along the timber bracing [27]. For all specimens, a significant deformation capacity could be observed.

Bending tests conducted on cross-halved (half-lapped) and mortise-and-tenon connections showed their moment capacity as shown in Figure 5.4. Cross-halved (CH) joints were tested in Japan with Japanese screw nails (2 pieces per each connection) and the same dimensions of specimens were also tested in Romania, with Romanian screws and also Romanian nails. The behavior of the later ones is half the ones in Japan. One argument can be that steel and timber are of much better quality in Japan. Can this be a tremendous influence on the behavior? Quality of materials, but most likely also the quality of workmanship!

Another interesting finding was that for bending, the screw types/lengths or nails, do not have a significant contribution to the capacity of the cross-halved connection. But only the timber's property of compression perpendicular to grain and friction.

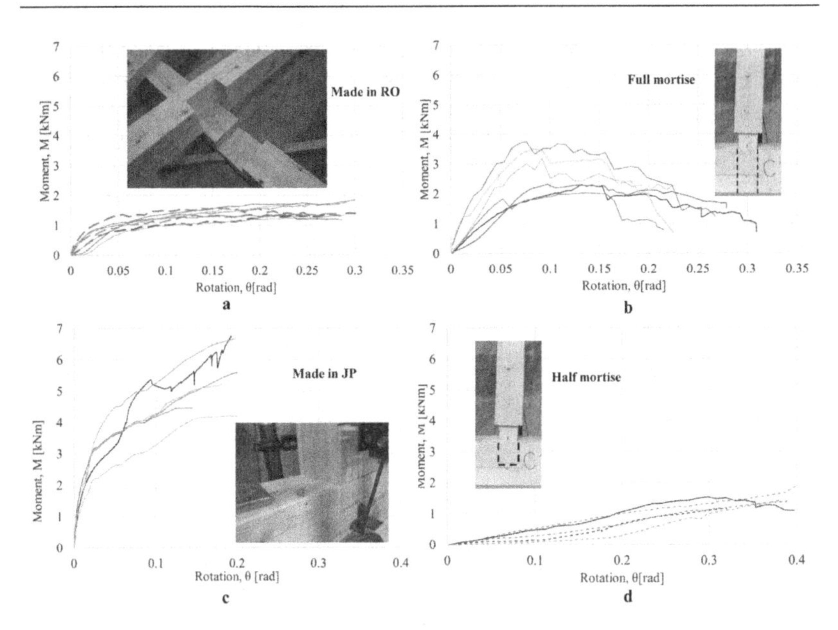

FIGURE 5.4 Moment-rotation capacity from bending test on connection for: (a) cross-halved (CH) connections s tested in Romania, having screws or nails; (b) mortise-and-tenon (M-T) with full penetration of the beam; (c) cross-halved (CH) connections with screws tested in Japan; (d) mortise-and-tenon (M-T) with half penetration of the beam.

Traction tests were also done on both mortise-and-tenon, and cross-halved connections and the results revealed that the workmanship (with no gaps) is very important and has a great impact on the behavior, especially for the mortise-and-tenon joint! Also, the longer the screws, the better, for the cross-halved connection. As for the cross-halved connection with nails, it was surprising that it has more stiffness than the ones with screws. We observed that the steel material of the screws is very brittle (in Romania), being easily ruptured during testing, and also their heads were easily broken during initial drilling.

Shear tests were also conducted to observe the cross-halved connections when subjected to a pure shear test. The results were as expected, only compression perpendicular to grain was observed, and no shear failure could be obtained [28].

Many people have a good opinion, based mostly on feeling rather on scientific facts, about the seismic behavior of *paianta* houses. As observed during earthquakes and experiments, which are still very few, the *paianta* house seems resilient when we define resilience as the capacity to withstand or to recover quickly from difficulties. They may suffer damage in strong earthquakes, usually the infills fall, but they can be reconstructed easily and even with the same material that has previously fallen. The timber frame would continue to support the roof loads even if the infills fall. The infills have a role in stiffening the timber frame, but mostly as a closure (envelope) of the house, to create a comfortable thermal environment inside the house. In some war areas, the infill is important to be strong enough to resist bullets (Pakistan – Kashmir). In some countries, the contact between the infill and the frame can be improved by adding nails on the timber frame (Portugal) or barbed wire in the masonry joints (Haiti), to prevent detachment in the out-of-plane direction.

One thing should be kept in mind, when we, engineers, state that *paianta* houses are resilient: that it depends on the frequency content of the earthquakes, and also on the interval during the main earthquake and big aftershocks. We are not sure if a *paianta* house would withstand a series (doublet or triplet) of earthquakes as the ones that recently catastrophically affected Turkey and Syria on February 6, 2023, where almost three cities were erased from the map, having all the buildings collapsed. This combination (two strong earthquakes within eight hours) and poorly constructed reinforced concrete buildings was never experienced before in our modern engineering history. Although, the Haiti earthquake in 2010 also demonstrated how deadly poorly executed reinforced concrete can be, and witnessed the timber frames with masonry infills (either *kai peyi* in the rural area or *Gingerbread* in the urban area) that were still standing, without killing its occupants, even though they experienced some damage.

New research ideas appeared after the Turkish disaster in Hatay, and all the buildings and standards must take into account the behavior in a sequence of several strong earthquakes. Shaking table tests for *paianta* houses can prove whether they can overcome that situation or not.

5.2.2 Japanese *Minka*

Among the literature, many experiments were done on traditional Japanese mud walls, many of which are focused on the hanging type of mud walls [29, 30]. The further experiment described was selected for the resemblance with the Romanian *paianta* experimental test, at least in the shape of the test specimens.

Dozo-style structures were mainly used by rich people or sellers and are part of the general traditional *minka* buildings. They are well maintained as they represent a valuable heritage for Japanese people. The function of *dozo*-style buildings is nowadays of stores/restaurants, but it can also be combined with residential function. They have earthen "mud walls" that are 200–300 mm thick at their outer circumference for fire protection. Although originally used

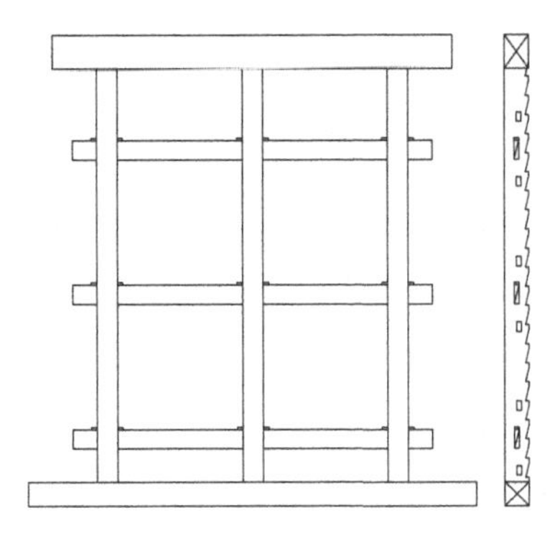
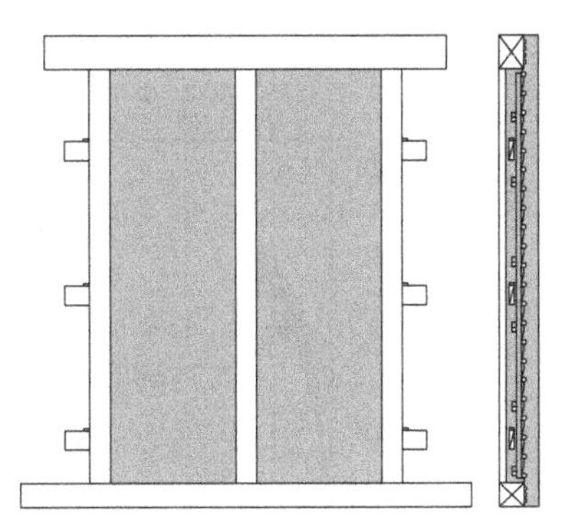

FIGURE 5.5 Tested specimens: Type A (left) and Type B and C (right) [31] (reproduced with the permission of Prof. Yokouchi and licensed under Creative Commons Attribution-NoDerivs (CC BY-ND)).

as warehouses, these structures came to be used as stores, parlors, and other kinds of buildings in modern times.

In [31] the construction and load testing of three types of specimens of *dozo*-style walls was done and further briefly presented. The type A specimen comprised the bare timber frame only, while the type B specimen had mud and bamboo elements infill and was used to clarify the strength, deformation performance, and damage state of the original mud wall of a *dozo*-style structure. The type C specimen had the same specifications as type B and was the test specimen used to clarify the effect of repairing walls damaged by earthquakes. The layout of the specimens is shown in Figure 5.5.

The shape and dimensions of the specimens and the materials used were basically the same: their width was two 910 mm spans, with a height of 2730 mm. The joint between the post and the horizontal frame was shaped so that the end of the post did not touch the horizontal frame, even in the case of significant deformation, thus only the performance of the wall panel could be observed. The shapes and dimensions of types B and C are shown in Figure 5.5.

The results of the wall tests are shown in Figure 5.6.

Bending tests were conducted by Professor Sakata et al. in [32] and [33] on typically used Japanese connections such as the mortise-and-tenon and timber dowel and also split wedge joint showed their resistance and they can be compared in Figure 5.7. The first type is mainly used for post-mudsills connections, while the second one mentioned is used for post and top plate or intermediary beams.

When comparing the Japanese types of connections, the traditional ones above, because there are also newer and much stronger types of connections, with the Romanian ones, it seems the Japanese ones are not as strong as the Romanian ones. But the tenon is usually thinner, and this may be the reason.

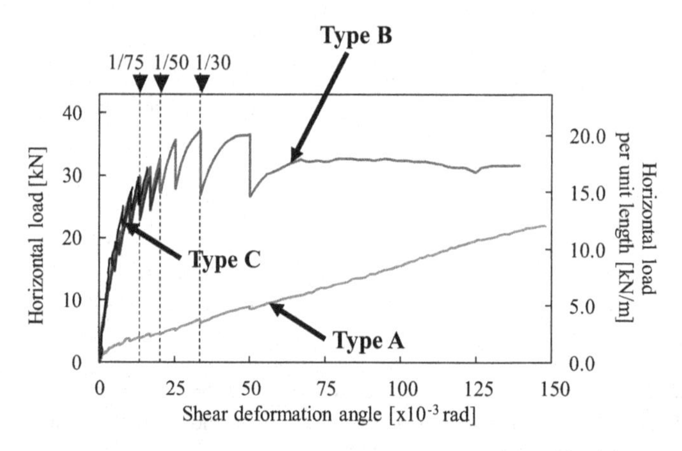

FIGURE 5.6 Envelope curves comparison on the positive loading [31] (reproduced with the permission of Prof. Yokouchi and licensed under Creative Commons Attribution-NoDerivs (CC BY-ND)).

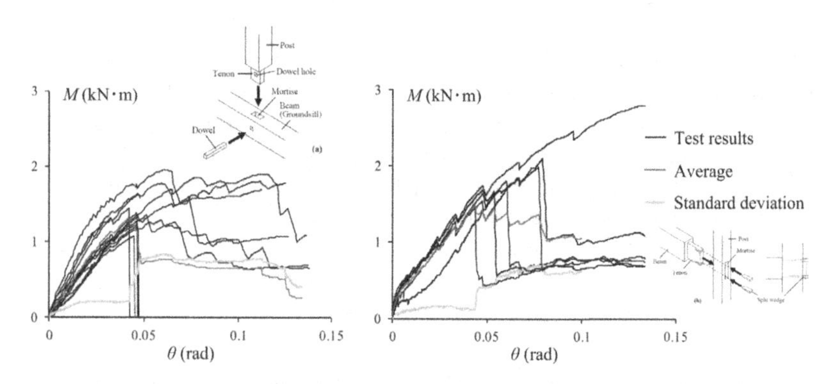

FIGURE 5.7 Moment capacity comparison of mortise-and-tenon (left) and split wedge connections (right) [33].

Other studies were done on other configurations of timber frames with infills, such as timber laths (10 mm thickness and 50 mm width) nailed to the columns and on which hemp rope was also attached to support a 20 mm layer of plaster, mixed with seaweed glue. This was a replacement of the mud infill, for western style buildings in Japan. Professor Matsumoto investigated experimentally such buildings, conducting wall tests in a cyclic static regime to identify the shear performance of such walls [34]. The hysteretic behavior and wall layout is shown in Figure 5.8.

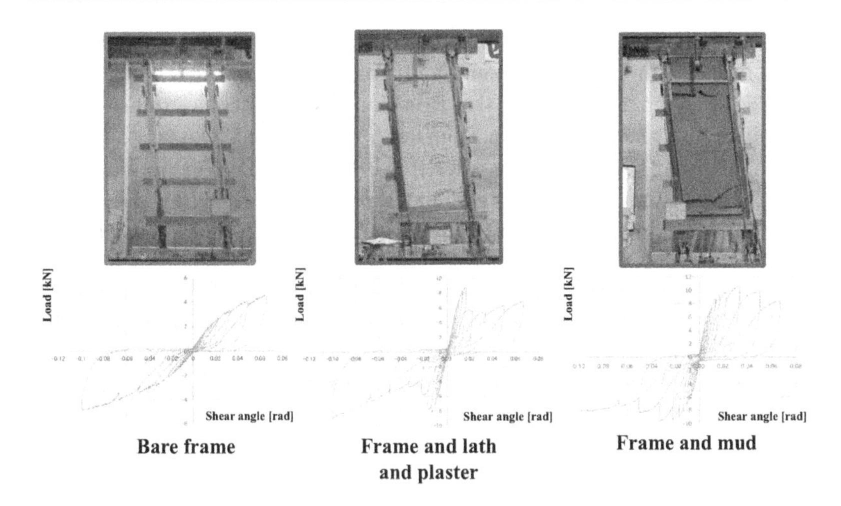

FIGURE 5.8 Experimental results on walls with timber frame (green line and border), timber frame and laths with plaster (red line and border) and timber frame and mud infill (blue line and border) [34] (reproduced with the permission of Prof. Matsumoto).

5.3 LOCAL SEISMIC CULTURE

The local seismic culture has a significant impact on the way people build their homes. Depending on the seismicity level (low, medium, or high), people are more willing to protect their houses.

Romania and Japan are two different seismic countries. While the first has strong earthquakes at longer intervals, the second one has them more often. So during one person's life, in Romania, it is possible that no seismic event will occur. But in Japan, each person experienced at least one strong earthquake, if not more. Thus, Japanese owners are more aware of the seismic resistance that their houses need to have, and consider it when buying/building a house, while in Romania, it is not so important as long as no recent earthquake occurred.

5.3.1 Romanian Seismicity

Romania is a country with moderate seismicity, and the major seismic events tend to occur with a recurrence interval of around 40 years. Thus, major earthquakes (that produced casualties and building collapses) happened in

1977, 1940, 1902, etc. Smaller magnitude earthquakes occurred after 1990, but not bigger than 6.4 Mw, without producing any negative effect, besides the reminder of the great population that Romania is still a seismic country.

During field investigation in Romania on traditional houses, one owner stated that she rebuilt her house after the 1940 earthquake using the type 1 of *paianta*, because she saw her neighbor's *paianta* surviving the earthquake without major damages, while her unreinforced brick masonry house didn't.

The seismic behavior of *paianta* is characterized as having a significant deformation capacity due to the timber frame, while the infills, no matter the type, give stiffness to the overall assembly. Due to the fragile fracture of the mud/clay mortar, even if combined with straws, the infills are prone to out-of-plane failures. But nevertheless, even if the infills fall, the timber structure does not usually collapse, and after the earthquake, the owner can rebuild the infills, sometimes without an intervention to the timber structure.

We could not see too much damage to *paianta* houses in the recent earthquakes also due to the nature of the earthquakes in Romania, which do not affect small period houses (with one or two stories). But the Vrancea seismic source needs more investigation, to understand the real nature of the earthquakes it produces.

The details of the connections in timber framing, although being done carefully in the past, with special attention to the perfect joint between the timber elements, without gaps, are not influencing that much the overall behavior, unless the infills are missing. The infills make a significant contribution, helping the timber frame to keep it in position when the earthquake occurs.

5.3.2 Japan

In Japan, the seismicity is characterized as maybe the highest in the world, earthquakes being a normal part of Japanese people's life. The occurrence of big events is also at small intervals, 5 to 15 years, when looking at the recent ones: Kumamoto 2016, Sendai 2011, Kobe 1995, etc. Other smaller ones occurred, one of 7.4 Mw even in 2022, again in Fukushima, producing four casualties and some damages occurring in the buildings.

But Japan invested in buildings research so much and for so long that they reached a level of buildings resistance that is hard to be equaled by other countries. Most of the technical solutions and design code recommendations are experimentally determined, based on a huge research work. All the calculations are easily explainable, and the design codes are easy to understand and apply.

New technology is easily adopted by Japanese owners, due to the imminence of the future big earthquake. Preparedness and awareness of the population is a priority, and also the alert system. Disaster drills are made yearly, and the emergency food is renewed. Both at home and at the office, citizens are taught about places to meet after a disaster, and name tags are written before the disaster drill so that when they arrive at the meeting point, they can easily be identified. In Tokyo Tech, helmets are also provided for all the employees.

Another thing worth mentioning is that the average lifetime of a building in Japan is 27 years, so the owners are careful about how to maintain or select their homes, considering the seismic behavior as one major thing to take into account when buying or making a building.

REFERENCES

[1] T. Makarios, and M. Demosthenous, "Seismic response of traditional buildings of Lefkas Island, Greece," *Engineering Structures*, vol. 28, no. 2, pp. 264–278, 2006. doi: 10.1016/j.engstruct.2005.08.002

[2] K. Cetin, A. Sextos, G. Mylonakis, and J. Stewart, "Seismological and engineering effects of the M 7.0 Samos Island (Aegean Sea) Earthquake," *Geotechnical Extreme Events Reconnaissance Association: Report GEER-069*, 2020.

[3] G. Koca, "Seismic resistance of traditional wooden buildings in Turkey," in *VI Convegno Internazionale*, Messina: Gangemi Editore International, 2018, p. 11.

[4] P. Gülkan, and R. Langenbach, "The earthquake resistance of traditional timber and masonry dwellings in Turkey," Paper no. 2297, 13th World Conference on Earthquake Engineering, 13th World Conference on Earthquake Engineering, August 1st–6th, 2004.

[5] S. Balan, V. Cristescu, and I. Cornea, *Cutremurul de pamant din Romania de la 4 martie 1977*. Bucharest: Editura Academiei Republicii Socialiste Romania, 1982.

[6] E. S. Georgescu, and M. M. Crainicescu, "Cladiri de locuit traditionale pentru zone rurale in Romania. Comportarea la cutremurul din 4 martie 1977," *Studii si cercetari INCERC*, no. 4, pp. 29–55, 1979.

[7] N. Ruggieri, "An Italian anti-seismic system of the 18th century decay, failure modes and conservation principles," *International Journal of Conservation Science*, vol. 7, no. 2, pp. 827–838, 2016.

[8] Z. Qu, A. Dutu, J. Zhong, and J. Sun, "Seismic damage to masonry-infilled timber houses in the 2013 M7.0 Lushan, China, earthquake," *Earthquake Spectra*, vol. 31, no. 3, 2015. doi: 10.1193/012914EQS023T

[9] R. K. Adhikari, and D. D'Ayala, 2015 Nepal earthquake: Seismic performance and post-earthquake reconstruction of stone in mud mortar masonry buildings, vol. 18, no. 8. Springer Netherlands, 2020. doi: 10.1007/s10518-020-00834-y

[10] A. C. Wijeyewickrema, K. Shakya, D. R. Pant, and M. Maharjan, "Cuee report 2011-1 earthquake reconnaissance survey in Nepal of the magnitude 6.9 Sikkim Earthquake of September 18, 2011," published by Tokyo Institute of Technology, Center for Urban Earthquake Engineering, 2011.

[11] K. M. O. Hicyilmaz, T. Wilcock, C. Izatt, J. da Silva, and R. Langenbach, "Seismic Performance of Dhajji Dewari," Proceedings of the 15th World Conference on Earthquake Engineering, Lisbon, Portugal, September 2012.

[12] S. Htwe Zaw, T. Ornthammarath, and N. Poovarodom, "Seismic reconnaissance and observed damage after the Mw 6.8, 24 August 2016 Chauk (Central Myanmar) Earthquake," *Journal of Earthquake Engineering*, vol. 23, no. 2, pp. 284–304, 2019. doi: 10.1080/13632469.2017.1323050

[13] R. Langenbach *et al.*, *Preserving Haiti's Gingerbread Houses: 2010 Earthquake Mission Report*, New York: World Monuments Fund, 2010. www.wmf.org/sites/default/files/article/pdfs/WMF%20Haiti%20Mission%20Report.pdf

[14] A. Dutu, K. Miyamoto, G. Jole Sechi, and S. Kishiki, "Timber framed masonry houses: Resilient or not?," in 3rd European Conference on Earthquake Engineering and Seismology (3ECEES), September, Bucharest, Romania, 2022.

[15] C. Cancino, S. Farneth, P. Garnier, J. V. Neumann, and F. Webster, "Damage assessment of historic Earthen buildings after the August 15, 2007 pisco, peru earthquake," pp. 1–68, 2011.

[16] B. Lizundia *et al.*, Eeri Ecuador 2016 earthquake reconnaissance team report, vol. 49, no. 11. 2016.

[17] M. Gallegos, and G. Saragoni, "Analysis of strong-motion Accelerograph records of the 16 April 2016 Mw 7.8 Muisne, Ecuador Earthquake," 16th World Conference on Earthquake Engineering, 16WCEE 2017, April 2016, pp. 1–12, 2017.

[18] A. Aldea, A. Dutu, S. Demetriu, and D. I. Dima, "Dynamic properties identification for a timber framed masonry house," *Constructii Journal*, vol. 1, pp. 3–12, 2020.

[19] N. Matsumoto, "Research on wall construction methods and structural performance of modern wooden buildings components of laminated plaster walls and horizontal force resistance mechanism," Ph.D thesis, University of Tokyo, 2016, doi: 10.15083/00075247

[20] M. Fujimoto, N. Matsumoto, and M. Koshihara, "The present state and issues on retrofitting of historic timber-framed brick construction buildings in Japan," *Curran Associates*, June 2023, pp. 4134–4140. doi: 10.52202/069179-0538

[21] A. Dutu, M. Niste, and I. Spatarelu, "In-Plane static cyclic tests on traditional Romanian Houses' walls," in 16 European Conference on Earthquake Engineering (16ECEE), Thessaloniki, Greece, June 18–21, 2018, pp. 1–12.

[22] A. Duţu, M. Niste, I. Spatarelu, D. I. Dima, and S. Kishiki, "Seismic evaluation of Romanian traditional buildings with timber frame and mud masonry infills by in-plane static cyclic tests," *Engineering Structures*, vol. 167, pp. 655–670, 2018. doi: 10.1016/j.engstruct.2018.02.062

[23] MDRAP, P100-1/2013. Cod de proiectare seismică – Partea I – Prevederi de proiectare pentru clădiri [Seismic Design Code – Part I: Design Prescriptions for Buildings]. Monitorul Oficial al României, No. 558 bis/3 September 2013 (in Romanian), 2019.

[24] A. Dutu, H. Sakata, Y. Yamazaki, and T. Shindo, "In-plane behavior of timber frames with masonry infills under static cyclic loading," *Journal of Structural Engineering (United States)*, vol. 142, no. 2, 2016. doi: 10.1061/(ASCE) ST.1943-541X.0001405

[25] A. Dutu, M. Niste, I. Spatarelu, and D. I. Dima, "TFMRO project (Seismic evaluation of Romanian traditional residential buildings)." [Online]. Available: http://tfmro.utcb.ro/

[26] V. Dragomir, *Conservation and Restauration of Traditional Architecture (in Romanian)*. Craiova, Romania: Editura Universitaria, 2012.

[27] R. Rupakhety and D. Gautam, Eds., Masonry Construction in Active Seismic Regions, 1st edition. Elsevier, 2021.

[28] A. Dutu, H. Sakata, and Y. Yamazaki, "Comparison between different types of connections and their influence on timber frames with masonry infill structures' seismic behavior," *16" World Conference on Earthquake, 16WCEE 2017*,paper ID 951, 2017.

[29] R. Ogushi, and H. Nakaji, "Experimental study on restoring force characteristics of Japanese traditional timber frames with strip-shaped mud-walls," The 17th World Conference on Earthquake Engineering no. 2, 2020.

[30] Z. Li, H. Isoda, A. Kitamori, T. Nakagawa, Y. Araki, and Z. Que, "Lateral resistance mechanism of a frame with deep beams and hanging mud walls in traditional Japanese residential houses," *Engineering Structures*, vol. 244, pp. 112744, 2021. doi: 10.1016/j.engstruct.2021.112744

[31] H. Yokouchi, and Y. Ohashi, "Repairing and recovering structural performance of Earthen walls used in Japanese dozo-style structures after seismic damage," *Journal of Disaster Research*, vol. 13, no. 7, pp. 1333–1344, 2018. doi: 10.20965/jdr.2018.p1333

[32] H. Sakata, T. Noguchi, Y. Yamazaki, and Y. Ohashi, "Experimental study on flexural-shear behavior of split wedge joint," *Journal of Structural and Construction Engineering (Transactions of AIJ)*, vol. 78, no. 690, pp. 1459–1467, 2013. doi: https://doi.org/10.3130/aijs.78.1459

[33] H. Sakata, Y. Yamazaki, and Y. Ohashi, "A study on moment resisting behavior of Mortise-Tenon joint with dowel or split wedge," 15th World Conference on Earthquake Engineering (15WCEE). Held in Lisbon, Portugal, 24th–28th September 2012.

[34] N. Matsumoto, K. Yamawaki, and K. Fujita, "Shear performance estimation of wooden laths and plaster walls through the experiment and analysis." World Conference on Timber Engineering Oslo, 2023.

6 Fire Resistance of Traditional Timber Frames

6.1 FIRE-RESISTANT SPECIES

There is a myth related to natural fire-resistant tree species, and in fact they have proven to be less flammable than other tree species [1], for example, the *Mediterranean cypress*, which due to its leaves and generally higher moisture content and also the protective properties of the bark, they do not catch fire in case of a wildfire. This means that the un-harvested tree has such qualities, but, for fire resistance as component members of a building, once the timber is dried, it will not retain these fire retardant properties.

6.2 TRADITIONAL FIRE RESISTANCE TREATMENT AND METHODS

Traditionally, no special method was applied on the elements of a traditional timber framed house to increase the fire resistance. The only method known as a retardant of fire is applying a layer of plaster made of a mix with clay and straw, and then a final layer of lime. But it was not the purpose of the application, the positive result was just accidental. In [2] the author states that the ASTM tests prove that the plaster applied (in that particular case on straw bales) represents an excellent fire retardant.

DOI:10.1201/9781003405375-6

6.3 MODERN FIRE RESISTANCE TREATMENT AND METHODS

In Romania, modern methods started being applied around the 1950s and the reference work is [3] which offers a very thorough description of the behavior in fire and the biological decay of timber, together with methods applied to increase the resistance of timber to such dangers.

It is first important to understand that when subjected to fire, timber burns, sooner or later. The later depends on us, since we have some ways to delay the fire collapse which basically occurs when the cross-section of the elements is weakened enough and loses its strength under its own weight.

While many textbooks present thoroughly the process, I will just briefly enumerate the steps of the fire behavior of wood, so the reader can understand in which way our current methods allow us to intervene and improve.

Up to 100°C, the moisture will evaporate. After that, the chemical decomposing of timber and hydrocarbons can form, ignite and burn. These gases can be ignited at a temperature around 230°C by an open flame, thus amplifying it, increasing and propagating the fire faster. If the temperature continues to rise above 350°C the carbon monoxide spontaneously ignites, and the wood surface starts to carbonize. Over 500°C the carbon monoxide quantity produced decreases, but the carbonization advances into the cross-section of the wood element. Due to the low thermal coefficient of timber, the carbonization is slow and the carbon layer actually protects the undamaged part of the cross-section from further burning. So when compared to other materials such as steel and aluminum, the timber elements keep their stability and fire resistance for a reasonable amount of time, just enough for people to evacuate safely their houses.

To improve the fire performance of a timber house there are several suggestions made by [4] and also found in other references [5] and these are:

- To provide means to extinguish the fire immediately like water supplies;
- Automatic sprinklers, which are known to be the most efficient method;
- Encapsulation which can be partial or complete, and it can consist of covering the timber elements with gypsum boards;
- Attention to connection details which often include metal fasteners, sensitive to fire;
- External fire spread barriers throughout the windows which can limit the fire propagation to the superior stories;
- Details to prevent the fire from spreading inside the house, such as internal fire stops, sprinklers, and smoke detection;

- Use of fire retardant substances applied on the timber elements by direct application or by void-pressure treatment.

The fire retardant substances which can be applied on the surface can be classified depending on their action modes and the final purpose, which are correlated to the burning process of wood, which was briefly described above. So one can use:

- Solid substances with a mechanical action that are applied as a liquid mix and provide an incombustible isolating solid cover that doesn't allow access to oxygen. Hereby the most common ones used for traditional timber buildings are plasters made of cement, lime, chalk, and ash;
- Solid substances which melt under heat and thus provide a protective layer;
- Silicate substances that through melting produce a foam cover;
- Substances (like ammonium salts) that during burning produce inert gases, which limit the access to oxygen;
- Substances that produce the charring of the contact surface (one example of this method is the sugi ban);
- Substances like ammonium phosphate or potassium carbonate that produce several protection actions.

The fire retardant substances must meet certain conditions for a successful application:

- To impregnate in the wood as much as possible, and this can be done best with the vacuum-pressure method;
- To be fixed well to the timber and to be weatherproof, this is done also with a previous cleaning of the surface;
- Not to attack in any way the support material (timber and sometimes metal fasteners);
- To be stable and effective for as long as possible;
- To be applied as easily as possible;
- To be cost-accessible;
- Not to favor the appearance of bio-decay.

There is no perfect solution for the treatment of wood against fire. If the method is very effective (vacuum-pressure impregnation), it is costly and complicated, because the elements must be fully immersed into industrial autoclaves. This cannot be done for an existing building, nor on the construction site of a low-rise residential house, since it is not cost-effective. Then, the surface application of the substances has only surface effectiveness, as it cannot reach the interior of the wooden element and it also has a limited time action, so after some years the treatment should be repeated.

The best way to protect a traditional house from fire remains the prevention and investment into prevention methods, like smoke detectors and sprinklers.

REFERENCES

[1] G. Della Rocca *et al.*, "Possible land management uses of common cypress to reduce wildfire initiation risk: A laboratory study," *Journal of Environmental Management*, vol. 159, pp. 68–77, 2015. doi: 10.1016/j.jenvman.2015.05.020

[2] B. King, *Building of Earth and Straw: Structural Design for Rammed Earth and Straw Bale Architecture*. Pacifica, California: Ecological Design Press, 1996.

[3] E. Vintila, *Protectia lemnului. Timber's Protection*. Bucharest: Editura Tehnica, 1959.

[4] A. Buchanan, B. Östman, A. Frangi, and A. Buchanan, "White paper on fire resistance of timber structures white paper on fire resistance of timber structures," 2014. doi: https://doi.org/10.6028/NIST.GCR.15-985

[5] C. Furdui, *Constructii din lemn: materiale si elemente de calcul*. Timisoara: Politehnica, 2005.

Strengthening Solutions 7

To understand the necessary strengthening solutions, first, it is needed to understand the causes and the type of damages. This chapter presents possible degradations of timber framed houses with infills specific to Romania and Japan, from both normal use and earthquake events. Detailed tables with all the possible damages classified into six categories are published in [1]. These categories of damage are accidental, environmental, mechanical, improper conformation, poor maintenance, and inappropriate interventions.

After classifying the damage and finding the actual cause, a strengthening solution can be proposed and Table 7.1 shows centralized possible damages associated to proposed strengthening and repair solutions.

7.1 DEGRADATIONS IDENTIFIED UNDER NORMAL USE (WITHOUT AN EARTHQUAKE)

The main problem of *paianta* houses remains the poor maintenance during the lifetime of the house. Photos with the most common degradations are shown in Figure 7.1. Many of the houses are abandoned, so the biggest enemy of the house is water infiltration.

While the water can come either from the rain/snow melting, it can also come from flooding of nearby rivers. As we already know, if the timber is wet/dry on a cyclic basis it will degrade fast, rotting beginning from the elements in contact with the water. This is the trigger to another enemy, the *biological decay* enhanced by fungus and by insects eating away the timber section.

Another problem is the *wrong conformation of the house*. Within this category, there are several situations. The most common one is *the lack of foundation* and for the poorest versions of houses, the mudsill sits directly on the soil. Usually, it sits on a brick masonry foundation, which has mud mortar as a binder, and it is usually cracked due to its drying properties. Sometimes the

DOI:10.1201/9781003405375-7

FIGURE 7.1 Possible structural damages in a paianta house.

foundation is made of river rocks, just placed one next to the other, without any binder; this is called dry masonry, and it is the best solution to cancel the capillary action of the water, but it is not a good technique for earthquake resistance.

Then, the *differential settlement* causes the house to crack (usually vertically in the walls). These cracks gradually increase if the foundation is not strengthened and/or the soil around stabilized. However, an interesting fact is that although the foundation might settle differentially, due to the configuration of the house, with timber posts at a maximum of 1.5 m in the longitudinal direction, the cracks propagation to the entire wall is limited (no cumulated cracks in the walls). Also, in some cases, even if the foundation is greatly inclined, the "box like" behavior of the upper timber frame usually prevents it from partial collapse. But in a few cases, when the house is completely abandoned, the house collapses without the occurrence of an earthquake.

The *connections* are usually flexible and sometimes present gaps due to poor workmanship, but in this configuration, such a mistake is not so important, as it doesn't usually lead to collapse.

The *presence of the diagonal braces* is an advantage to the good performance of the house, although the reason they are used is usually purely for construction support, while building the posts and beams. But their layout, if *not symmetrical*, may introduce different torsion or behavior to the whole house. Sometimes they are not connected to both of the posts, but just to one of the posts, and at the other end they are connected to the beam only.

The *infill* is usually sensitive to out-of-plane actions, as it is *not connected in any way to the timber frame.* Damage is more commonly caused by earthquake, not due to poor maintenance. *Plaster falling*, especially at the contact between the timber frame and the infill, is not a structural problem, but the cause may in time affect the structure also.

7.2 DAMAGES DUE TO EARTHQUAKES

Damages in earthquakes to timber frames with infills were observed in the Haitian earthquakes (2010 and 2021) and they consisted mainly in out-of-plane falling of the infills. The same kind of damage was observed in the Lefkada earthquake, and it is said that Greek people know that when an earthquake occurs and they live in such a house, they shouldn't go out of the house, but stay inside, because the infills usually fall outside the house.

Plasters often crack and fall, especially at the corners of the walls, at the contact between the timber frame and the infills.

In the Japanese case, the traditional houses are sometimes damaged in earthquakes due to the high dead load of the roofs made of ceramic tiles, and in the case of subsequent earthquakes, if the house leaned in the first one, it is prone to collapse in the next big one if it is not strengthened. Thus connections, especially at the bottom between the mudsill and the posts, fracture and may produce the collapse of the entire house.

7.3 COMPATIBLE STRENGTHENING SOLUTIONS

One thing must be initially mentioned here, as we are focusing on the traditional houses that are not registered as national heritage monuments, which have a specific regime when considering their strengthening, and is not the same thing as a restoration, mandatory for such a specially classified building. Thus, if a heritage building restoration is necessary, this means that, based on the Charta from Venice, adopted in 1964, the same construction techniques must be used, and the objective is to bring back the building to its original capacity. But in this book, we are only referring to the general traditional houses, not included in the national heritage monuments list, for which the owners can just apply any strengthening technique that is available without the need to ask for a special permit, other than a construction approval.

In Romania, the national buildings standards do not cover the traditional timber framed houses, and thus, the owners do not actually know how to strengthen the houses or how to improve their thermal comfort such as the new materials to be compatible with the original ones. For this reason, in many situations concrete or cement-based mortar/finishing reinforced with steel-welded mesh is applied and, on the walls, even if structurally efficient is not compatible with the earth or lime mortar within the infill. Moreover, even structurally, the concrete may cancel the ductile behavior of the timber frame and introduce a brittle failure mechanism due to the resistance of the concrete cover added. Or, in an earthquake, the cover can just detach in the initial cycles and have parallel behavior, not helping but just adding an additional burden on the original timber frame of the wall which may lead to faster damage of the frame.

In Japan, there is another story about strengthening techniques, as the people here are used to frequent earthquakes, and appreciate, accept and pay happily for new and efficient technology for structural improvement of traditional houses, such as visco-elastic dampers or even BRBs as diagonals to limit the displacement of the walls and guide the energy dissipation not mainly through the connections, but by the passive control devices installed. Also, for the national heritage timber frames with infills, specialists conduct extensive research to find the most appropriate strengthening solution. One such example is the Tomioka Silk Mill, for which shaking table tests were conducted to understand how it behaves. The results were very interesting, and show that the last layer of mortar, between the upper part of the infill and the top plate (upper timber beam) has a great influence on the out-of-plane behavior of the wall, even if apparently the mortar and timber do not really attach to each other. The friction between them does enough work to overcome the collapse of the infill in the out-of-plane direction of the wall. Another finding was that the aramid fiber rods inserted in the joints (also called the near surface mounting strengthening technique) were successful in confining and adding enough stiffness and shear capacity to the infills, and connecting them to the frames, so they could withstand the Japanese earthquakes.

Further, several strengthening techniques will be enumerated and references to the experimental studies proving their effectiveness will also be provided:

a. TRAROM solution [2], which was applied also for the strengthening of type 1 *paianta* houses, and the experimental results showed that the infill does not contribute as much as the diagonal planks applied on both faces of the strengthened wall;
b. Improving the connections with metal fasteners [3, 4] is always an option to increase the stiffness of the timber frame. From experiments on a timber frame with and without metal plates applied only on the connections of the diagonals, the maximum capacity of the frame increased from 5.2 kN to 12.3 kN with the metal plates;
c. Seismic bands [5] work well when the masonry infill panels are wider than 1 m^2 and even on the triangular ones of the type 1 *paianta* house;
d. FRP bands, textiles or rods [6] represent a possible solution, but kind of expensive for a residential house which is not heritage. Textiles or

laminates can be attached mainly to the timber connections, taking care of the irregular surfaces or the difference in the cross-sections of the connected timber elements. Near surface mounting of rods is also possible and proved to be effective [7].

e. Meshes either from polymers or welded steel [8]–[11], used with a lime or mud mortar for the first type, and cement mortar for the later are also an option. The cement version is already widely used on the *paianta* walls, but many specialists think that cement mortar is not compatible with the earth from the walls' infills, also due to the different stiffness of the materials, which in the case of an earthquake will separate and basically work as two different elements, and this does not bring any improvement to the wall, in fact;

f. Improvement of the connection (adherence) between the timber and masonry [12] by inserting nails on the contour of the masonry infills, or some barbed wire in the joints, but this latter one works for newly built or reconstructed infill;

g. PP (polypropylene) bands [13]–[15] are also offering a confining effect to the overall wall, an increased strength and ductility, while one important challenge being the connection between the PP bands. This should be done with a special ultrasonic welding machine for plastic, but it is not easy to buy or to borrow one.

h. Tie rods applied on the vertical direction, at 20 cm from the posts on the horizontal direction, which prevent the deformation of the timber frame and for large deformations they are tensioning, increasing the lateral strength of the masonry, by adding axial force on it. This strengthening solution works for brick masonry infills and has the feature of bringing the wall back to its original stiffness and strength, nothing more and nothing less.

i. NSM (near surface mounted) strengthening with rods inserted in the masonry joints, works well for brick masonry infill, and the rods can be made of composites, especially when the original mortar is made with materials incompatible with steel rods [16, 17].

j. Seismic isolation can be a solution, but maybe more suitable for heritage timber frames with infills, and usually it also needs a reinforcing strong structure underneath which the isolation interface can be executed. Low-cost seismic isolation are also being studied and applied recently, having either a PVC rollers layer [18] or concrete filled tennis balls [19, 20] as the isolation interface, or also sand on Teflon sheet. The objective being that the superstructure would just slide, being decoupled from the foundation.

k. A steel supporting frame is also usually applied in Japan as a strengthening technique, transferring the loads to this strong frame,

TABLE 7.1 Strengthening solutions proposed based on the cause and effect of the degradation

	CAUSE	*EFFECT*	*STRENGTHENING SOLUTION*
Accidentally Produced Degradations	Landslides/soil creep causing differential settlements	Vertical and inclined cracks	Soil stabilization, cracks' injection and strengthening of the connections
		Out-of-plane deformations	Pushing back to initial position of the deformed element and strengthening the connections together with appliance of a wire mesh on the infills to prevent the out-of-plane failure
	Earthquakes	Out-of-plane deformation	Pushing back to initial position of the deformed element and strengthening the connections together with appliance of a wire mesh on the infills to prevent the out-of-plane failure
		X-shaped cracks	Cracks' injection and application of wire meshes
		Joints detachment	Strengthening of the connections
	Floods	Enhancing biological decay	Dry the house, apply chemical substances
		Degradation of the socle masonry	Repair the masonry with compatible solutions and dig a trench and add a drainage system

(*continued*)

TABLE 7.1 (Continued)

	CAUSE	*EFFECT*	*STRENGTHENING SOLUTION*
	Storms	Detachment of the roof cover	Replace the roof cover
		Detachment of the rainwater drainage elements	Reattachment of the undamaged drainage elements and replacing of the damaged ones
	Fire	Total or partial destruction of the timber elements	Reconstruction
Environmentally-induced degradations	Heavy precipitation/ stagnating water near the walls/ water washing the facade of the house	Humidity from the capillary action causing:	Extending the eaves, installing drainage systems around the house
	Heavy precipitation/ stagnating water near the walls/ water washing the facade of the house	degradation of the socle masonry – mold, fungus, algae, bacteria proliferation, – which finally results in:	Extending the eaves, installing drainage systems around the house Cracks' injection
	Freeze-thaw cycles, thermal shocks due to rapidly changing temperature	biological decay of mudsills and wooden facade cladding – plaster peeling	

	– exposure of wall structure to environmental factors – cracks	
Freeze-thaw cycles, thermal shocks due to rapidly changing temperature	Spalling	Reconstruction of the finishing layer increasing the adherence between the support layer and the new finishing
Deposits of green algae of lichens on the surface of exterior finishes	Deformations of the timber structure due to the sudden loss of moisture content, the increase of humidity relative to the decrease in temperature	Replacing the elements that lost their stability and strengthening the connections
	Degradation of timber elements	Cleaning the timber elements with chemical solutions and replacing the parts which lost too much of their mass
Rodents and bird droppings	Reduction of timber elements cross-section	Cleaning the timber elements with chemical solutions and replacing the parts which lost too much of their mass
Rodents and bird droppings	Harmful bacteria proliferation	

(*continued*)

TABLE 7.1 (Continued)

	CAUSE	*EFFECT*	*STRENGTHENING SOLUTION*
Mechanically-induced degradations	Poor foundation soil	Differential settlements, out-of-plane deformations	Soil stabilization around the house, cracks' injection and strengthening of the connections
	Undrained soil	Increased capillary action and humidity	Installing drainage system around the house (at the bottom)
	Collapsible soil	Differential settlements	Soil stabilization around the house
Degradations caused by improper conformation	Lack of foundations/ insufficient foundation depth	Detachment of timber elements	Strengthening of foundations
		Cracking	Cracks injections
		Differential settlements	Subfoundation
	Supplementary loading caused by changes in functionality or by the application of heavy finishings	Cracking	Injections with lime/mud mortar
		Detachment of timber elements	Strengthening of the timber frame's connections with metal plates and screw nails
		Loss of strength and stability of load-bearing walls	Additional supports (posts or braces)
	Load pattern errors, inappropriate/ insufficient load-bearing elements	Cracks	
		Timber elements bending	Additional supports (posts or braces)
	Poorly designed/ executed joints	Joints detachment	Strengthening of the timber frame's connections with metal plates and screw nails

Timber defects	Crooked timber elements	Not necessary
	Holes in timber elements	Injection with resins or lime mortars
	Cracks in timber elements	If they are superficial, no intervention is necessary. If they are deep, the element can be wrapped in some textile material or steel.
Use of already infested timber for construction	Timber disintegration	Replacing of the affected part
Undried timber with over 30% humidity	Crooked/shrinked timber elements	Not necessary
	Cracked timber elements	If they are superficial, no intervention is necessary. If they are deep, the element can be wrapped in some textile material or steel.
Use of softwood timber	Fast biological decay due to low durability of the timber species	Replacing of the affected part
Socle made of highly hygroscopic materials	Deterioration of the socle masonry	Repairing of the damaged part
	Loss of foundation strength	Replacing the foundation on small sections, not to affect the stability of the whole house
	Biological decay of mudsills	Replacing of the affected part.

(*continued*)

TABLE 7.1 (Continued)

	CAUSE	*EFFECT*	*STRENGTHENING SOLUTION*
		Superstructure dislodgement/ sliding	Supporting the structure and replacing the affected parts
	No perimeter pavement around the house	Moss, algae or lichen films appear at the base of the walls	Digging a trench and inserting a drainpipe or executing an inclined pavement
	Insufficient eave overhangs (widths)	Socle and walls exposed to various types of degradation due to excessive humidity exposure	Adding overhang to extend the eaves
	North-facing facades exposed to high humidity without the possibility of drying	Biological decay on north-facing facades	Adding overhang to extend the eaves
Degradations caused by poor maintenance	Failing to timely intervene for eradicating a biological attack of the facade	Biological decay	Replacing of the affected parts

Allowing abundant vegetation and trees to shade facades, thus maintaining increased facade moisture, roots penetrating the houses' foundation		Clean the house's surrounding, repair the foundations
Uninhabited house, unheated, producing inside an increase of the air humidity		Ventilate and heat
	Cracks in the clay mortar joints	Injections with resins or mortar (lime or mud)
Lack or degradation of the rainwater drainage system—gutters and downspouts	Permanent humidity in the area of the socle and the facades (rain washing the facades), favoring the degradation of the plaster and then of the infill and timber structure	Add/Repair/Replace the gutters and downspouts system
	Destruction of the eaves	Repair the eaves

(*continued*)

TABLE 7.1 (Continued)

	CAUSE	*EFFECT*	*STRENGTHENING SOLUTION*
	Lack of windows and doors	Increase of the relative humidity indoors, allowing the penetration of insects, birds, people, etc., hence biological degradations as a consequence of destructions caused by humans	Add the missing windows and doors
Degradations caused by inappropriate interventions	Embedding original socle in concrete, attempting its retrofitting	Capillary action	Add chemical barrier between the socle and the mudsill
		Biological decay	Replace the affected parts
	Improper joint strengthening	Detachment of timber elements	Remove the bad original solution and add metal plates to strengthen the connections
	Additional elements or bearings changing the static scheme of the structure	Bent timber elements	Reconsider the static scheme of the house and add support elements
		Cracks	Reconsider the static scheme of the house and inject the cracks with resins or mortar (mud or lime mortar)

Cement plastering on the socle and walls	Biological decay	Add chemical barrier between the socle and the mudsill and replace the damaged elements/parts
	Plaster spalling	Add chemical barrier between the socle and the mudsill, dry and replace the damaged parts
	Spalling of the masonry along with the cement plaster	Add chemical barrier between the socle and the mudsill, dry and replace the damaged parts
Use of asphalt board waterproofing over timber elements	Mudsill biological decay	Remove the board and replace the damaged parts
Mudsill placed directly on the soil or on a brick masonry foundation	Biological decay	Add chemical barrier between the soil/socle and the mudsill
Lack of socle	Biological decay of timber elements	Subfundation and replace the damaged elements
Poorly designed interventions to low-durability mudsills	Biological decay of timber elements	Replace the damaged elements
Lack of traditional construction knowledge	Structural instability	Reconsider the static scheme of the house and add support elements

(*continued*)

TABLE 7.1 (Continued)

	CAUSE	EFFECT	STRENGTHENING SOLUTION
	Lack of knowledge on modern materials compatibility with natural, traditional ones	Biological decay	Replace the damaged elements
	Use of vapor-proof varnishes or paints that alter the wood structure	Biological decay	Remove the varnishes and replace the damaged elements
	Adding waterproof thermal insulation	Biological decay	Remove the incompatible thermal insulation and replace the affected elements, add a compatible insulation
	Building concrete walkways or paving tangent to the socle	Enhancement of capillary action, dampness	Make a slit between them
	Existence of heavy machinery traffic roads in the vicinity of the building (inducing vibrations into the house structure)	Wall cracking	Injections with resins or mortars (mud or lime)
		Differential settlements	Stabilize the soil, strengthen the foundation and repair the house's structure

while keeping the original timber frames just as a façade system, or for aesthetical reasons [16].

As guidance depending on the problem an owner encounters when living in a timber frame with infills house, no matter the location, the following Table 7.1 provides a wide view on the strengthening solutions one can apply depending on the cause and the effect. When choosing the appropriate strengthening solution, it is compulsory to look first at the cause, solve that one and then treat the effect. Otherwise, if just treating the effect, which is in fact the degradation, the problem may reappear soon after the house is repaired or strengthened. Table 7.1 gives the strengthening solutions as if the cause has been solved already.

REFERENCES

[1] A. Dutu, M. Niste, I. Craifaleanu, and M. Gingirof, "Construction techniques and detailing for Romanian Paiant ă Houses: An engineering perspective," *Sustainability (Switzerland)*, vol. 15, p. 1344, 2023. https://doi.org/10.3390/su15021344

[2] A. Dutu, D. Barbu-Mocanescu, M. Niste, I. Spatarelu, and Y. Yamazaki, "In-plane static tests on a structural timber frame system proposal (TRAROM) inspired from traditional architecture and using local materials," *Engineering Structures*, vol. 212, no. March, p. 110491, 2020. doi: 10.1016/j.engstruct.2020.110491

[3] A. Dutu, H. Sakata, and Y. Yamazaki, "Comparison between different types of connections and their influence on timber frames with masonry infill structures' seismic behavior," 16th World Conference on Earthquake, 16WCEE 2017, vol. 951, no. Abstract ID, 2017.

[4] T. Schacher, and Q. Ali, *Dhajji Construction: For One and Two Storey Earthquake Resistant Houses – A Guidebook for Technicians and Artisans*, cantons Ticino, Valais and Graubünden: University of Applied Sciences and Arts of Southern Switzerland (SUPSI), 2009.

[5] S. Yadav *et al.*, "Effects of horizontal seismic band on seismic response in masonry structure: Application of DIC technique," *Progress in Disaster Science*, vol. 10, 2021. doi: 10.1016/j.pdisas.2021.100149

[6] A. Dutu, H. Sakata, Y. Yamazaki, and T. Shindo, "Influence of an AFRP retrofit solution when applied to timber framed masonry panels," in WCTE 2016 – World Conference on Timber Engineering, Wien, 2016.

[7] E. Poletti, G. Vasconcelos, J. M. Branco, and A. M. Koukouviki, "Performance evaluation of traditional timber joints under cyclic loading and their influence on the seismic response of timber frame structures," *Construction and Building Materials*, vol. 127, pp. 321–334, 2016. doi: 10.1016/j.conbuildmat.2016.09.122

[8] D. Torrealva, J. Neumann, and M. Blondet, "Earthquake resistant design criteria and testing of adobe buildings at Pontificia Universidad Catolica del Peru," *Proceedings of the Getty Seismic Adobe*, Part (volume) 1, pp. 3–10, 2006.

[9] F. Peña Mondragón, and P. B. Lourenço, "Criterios Para El Refuerzo Antisísmico De Estructuras," *Revista de Ingeniería Sísmica*, vol. 66, no. 87, pp. 47–66, 2012, doi: 10.18867/ris.87.45.

[10] T. L. G. Michiels, "Seismic retrofitting techniques for historic adobe buildings," *International Journal of Architectural Heritage*, vol. 9, no. 8, pp. 1059–1068, 2015. doi: 10.1080/15583058.2014.924604

[11] H. Yokouchi, and Y. Ohashi, "Repairing and recovering structural performance of Earthen walls used in Japanese dozo-style structures after seismic damage," *Journal of Disaster Research*, vol. 13, no. 7, pp. 1333–1344, 2018. doi: 10.20965/jdr.2018.p1333

[12] A. Dutu, J. G. Ferreira, A. M. Goncalves, A. Covaleov, and P. Fellow, "Components cooperation in timbered masonry," *Constructii*, vol. 1, pp. 1–6, 2012.

[13] N. Sathiparan, "State of art review on PP-band retrofitting for masonry structures," *Innovative Infrastructure Solutions*, vol. 5, no. 2, pp. 1–18, 2020. doi: 10.1007/s41062-020-00316-9

[14] N. Sathiparan, and K. Meguro, "Shear and flexural bending strength of masonry wall retrofitted using PP-band mesh," *Constructii*, vol. 14, no. 1, pp. 3–12, 2013.

[15] Y. Shen, X. Yan, H. Liu, G. Wu, and W. He, "Enhancing the in-plane behavior of a hybrid timber frame–mud and stone infill wall using PP Band Mesh on one side," *Polymers*, vol. 14, no. 4, 2022. doi: 10.3390/polym14040773

[16] M. Fujimoto, N. Matsumoto, and M. Koshihara, "The present state and issues on retrofitting of historic timber-framed brick construction buildings in Japan," *Curran Associates, Inc.*, June 2023, pp. 4134–4140. doi: 10.52202/069179-0538

[17] T. Hanazato, Y. Tominaga, T. Mikoshiba, and Y. Niitsu, "Shaking Table Test of Full Scale Model of Timber Framed Brick Masonry Walls for Structural Restoration of Tomioka Silk Mill, Registered as a Tentative World Cultural Heritage in Japan," in *Historical Earthquake-Resistant Timber Frames in the Mediterranean Area*, N. Ruggieri and G. Tampone, Eds., Switzerland: Springer International Publishing, 2015, pp. 83–89.

[18] A. Tsiavos, A. Sextos, A. Stavridis, M. Dietz, L. Dihoru, and N. A. Alexander, "Experimental investigation of a highly efficient, low-cost PVC-Rollers Sandwich (PVC-RS) seismic isolation," *Structures*, vol. 33, pp. 1590–1602, 2021. doi: 10.1016/j.istruc.2021.05.040

[19] A. A. Katsamakas, and M. Vassiliou, "Lateral cyclic response of deformable rolling seismic isolators for low-income countries Lateral cyclic response of deformable rolling seismic isolators for low- income countries," in *3ECEES Bucharest*, 2022. doi: 10.3929/ethz-b-000570039

[20] A. A. Katsamakas, M. Chollet, S. Eyyi, and M. F. Vassiliou, "Feasibility study on re-using tennis balls as seismic isolation bearings," *Front Built Environ*, vol. 7, 2021. doi: 10.3389/fbuil.2021.768303

Conclusions 8

In Romania a few specialists (architects and engineers) are fighting to keep the old traditional houses, teach people how to improve them without destroying them and value them, in spite of the general tendency to demolish them. In most parts of the world, traditional construction is not regulated through design codes. But in Japan, there is a code dedicated just to traditional residential construction methods, based mostly on experimental research, and engineers can easily follow the design recommendations, when designing a new house even if it has a traditional architecture. In this way, cultural identity is kept and also the safety of the owners is achieved.

It is interesting to observe that Japan is more open to experimental research than Europe, including Romania. Besides the fact that they have a design manual for traditional construction, they built it based on extensive experimental testing on walls and connections.

In the Japanese code, each vertical element of the structure (wall or post, usually including joints) that contributes to the lateral force-resisting system is considered a force-displacement curve, simplified to multi-linear (elasto-plastic). They are summed together and thus the total curve of the house in one direction is determined.

The Romanian way is to calculate the capacity against the requirement for each contributing individual element (connection, post, beam, etc.), probably due to the lack of experimental data on walls. This way, the estimation is less accurate, amplified also by the fact that the infill and connections collaboration within a wall is not considered.

An interesting difference between the two types of houses presented in this book is the lack of diagonals in Japanese traditional houses. Although Japanese specialists know the benefit of the bracing elements (i.e. they invented the buckling restrained braces – BRBs – and they are now testing and considering their applications in tall timber buildings), the traditional way is without diagonals. Thus, unlike in Romania (and probably Europe), the local seismic culture could not overcome the traditional way of building houses.

DOI:10.1201/9781003405375-8

Step by step, learning day by day, Romania is also making steps towards a national building code on traditional houses, thus legalizing them and familiarizing engineers and architects on appropriate construction details, which should meet the current needs in terms of safety, comfort, and durability. Due to the lack of research funding allocated at the national level, experiments are scarce on traditional houses' sub-assemblies, but given the already implemented research fundings (TFMRO and STRONGPa), we are closer to our goal to understand more about their behavior under different loadings.

Although many parameters are not yet having equations to be estimated (at least in Romania), this is an ongoing research topic, and it can grow easily if more engineers would focus on it.

Looking at the *minka* and *paianta* houses, from a bird's eye view, we notice that the construction details are not very different, the materials used being the same or in a similar way. The clay is used as an infill, timber elements as the skeleton forming the post-and-beam structure, with and without braces. The mortise-and-tenon type of joint is common to both types of houses. But *minka*, due to the exceptional carpentry tradition in Japan which is also well kept and transmitted to the next generations, may have more complex joints.

This book was meant to introduce the reader to the engineering details of traditional houses in Romania and Japan, namely the *paianta* and *minka* houses, respectively. It gives an idea of how to build such a house, assuming the reader may want to do it himself/herself. And it also shows experimental results and earthquake proofs on how such houses may behave in case of a seismic event. In the end, design examples are added, giving the reader a starting point on how to judge the design method of such a house.

Annex 1
The Design of a *Paianta* House

A1.1 INTRODUCTION

This calculation is just an exercise, taking as a reference the dimensions of existing houses, but assuming that we are designing and verifying it according to the current building codes.

The layout of the house's walls is shown in Figure A1.1, it has only one story (the ground floor) with the total dimensions of 4.65 m by 12 m. The typology is the classic *paianta* house with fired clay brick masonry infill (type 1 – according to Chapter 3).

The location of the house is in Prahova County, Tufani village, Romania. The roof is of the hip type.

A rigid-diaphragm flooring system is assumed (*assumption 1*). The columns at the base and the roof have the translations constrained in all directions (*assumption 2*).

The calculations do not include the foundations, which, according to the current codes should be made of reinforced concrete, and the connections between the foundations and mudsills, are also not included. They can be made with hold-down connections (as recommended in Chapter 3) and thus can be strong enough to carry the loads coming from the superstructure to the foundations. Moreover, in many existing houses, the mudsill is simply supported on the socle or foundations (if existing).

The design procedure shown in this chapter is simplified, and only linear static analysis is done. The timber elements are drawn using SAP2000

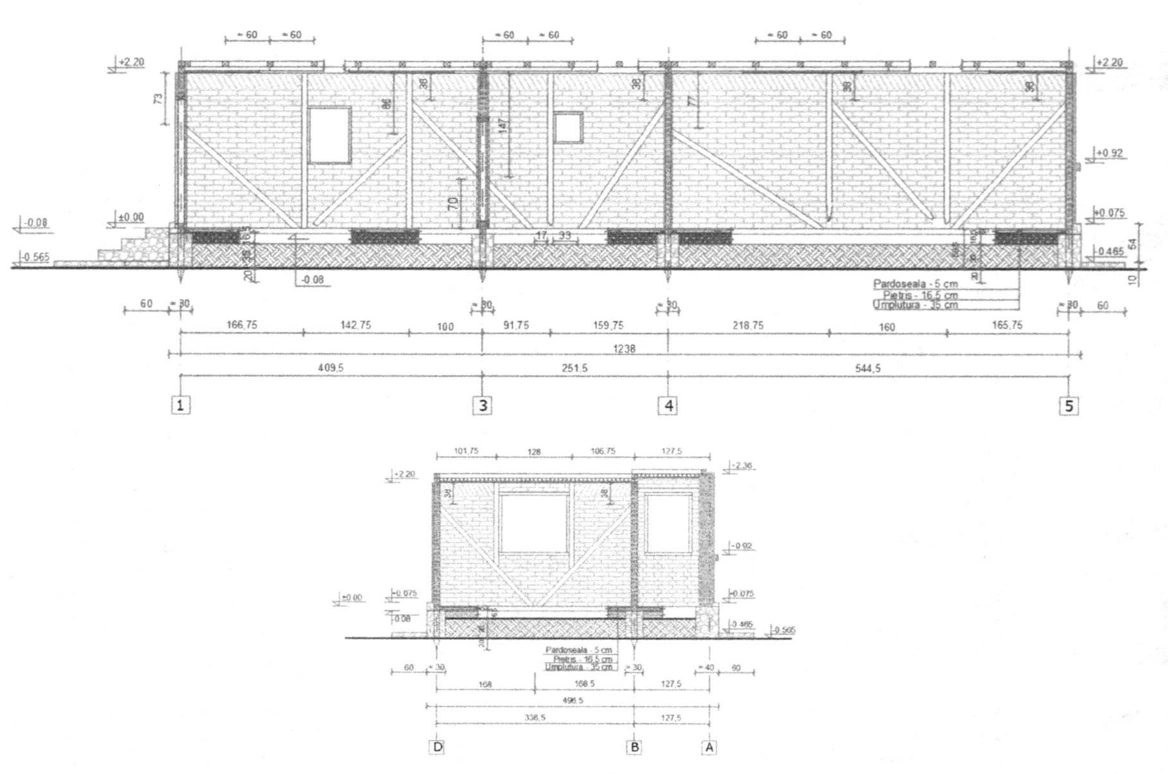

FIGURE A1.1 The *paianta* house's post layout in a longitudinal direction (up) and in a transverse direction (down).

TABLE A1.1 Characteristic material properties associated with the used timber class (C16)

MATERIAL PROPERTY	UNIT	VALUE
Bending strength	$f_{m,k}$ (MPa)	**60.59**/66
Tension strength in the fiber direction	$f_{t,0,k}$ (MPa)	90
Tension strength perpendicular to the fiber direction	$f_{t,90,k}$ (MPa)	2.7
Compression strength in the fiber direction	$f_{c,0,k}$ (MPa)	40
Compression strength perpendicular to the fiber direction	$f_{c,90,k}$ (MPa)	5.8
Shear strength	$f_{v,k}$ (MPa)	6.7
Rolling shear	$f_{T,k}$ (MPa)	4
Characteristic modulus of elasticity parallel to the grain	$E_{0,k}$ (MPa)	**8900**
Shear modulus	G_{mean} (MPa)	500 (for C16)
Characteristic density of timber	r_k (kg/m^3)	**385**
Average moisture content of timber	%	**15**
Characteristic density of clay (earth)	r_c (kg/m^3)	**1600**

commercial software, loaded as given from the calculations (seismic, wind, and snow), and with the resulting forces in the elements, design verifications are made.

The timber species considered is the Romanian fir and properties are based on the experiments conducted and described by [1] are bolded in Table A1.1. The rest of the values are considered based on the ones given in [2].

The calculations are made taking as reference the Eurocode 5 and the general formulas for each parameter are as follows:

$$X_d = k_{mod} \times X_k / \gamma_M \quad \text{(Eq. 1)}$$

$$R_d = k_{mod} \times R_k / \gamma_M \quad \text{(Eq. 2)}$$

X_d represents any design parameter, while X_k is any characteristic value for the considered parameter.

γ_M is the partial safety coefficient, taken as 1.0 hereby (according to [2]);

k_{mod} is a load bearing modification factor, depending on the loading type and the importance class of the building as in Table 4.11 according to [2]). Taken hereby as 1.1 corresponding to the seismic scenarios for exposure class

1 [3]. The exposure classes for timber buildings refer to the climatic conditions where the building is located: the ambient temperature, relative ambient moisture content and average moisture content of timber [3].

A1.2 LOADS CALCULATIONS DEPENDING ON THE LOCATION OF THE HOUSE

A1.2.1 Dead Load

A1.2.1.1 Roof load

The roof load comes from the roof cover and framework and is calculated as follows:

$$g_{framework} = V_{roof\ framework} \cdot \rho_k = 1.35\ \text{m}^3 \cdot 3.8\ \text{kN}/\text{m}^3 = 5.1\ \text{kN}$$

where $V_{roof\ framework}$ is the volume of the timber elements forming the roof framework, considering a cross-section of 120 x 120 mm for all the elements and ρ_k is the specific weight of the timber, according to Table A1.1. This also includes the deck applied on the rafters, to support the roof cover.

$$g_{roofcover} = A_{roof\ cover} \cdot \rho_{ceramic\ tile} = \left(60\ \text{m}^2 / \cos 35°\right) \cdot 0.4\ \text{kN}/\text{m}^2 = 29\ \text{kN}$$

where $A_{roof\ cover}$ is the area covered by the ceramic tiles, which is the horizontal area of the roof multiplied by cos (α), with α taken as 35° and $\rho_{ceramic\ tile}$ is the specific weight of the used tiles.

A1.2.1.2 Slabs

The slab considered is the one at the roof level. Considering that the roof slab is made of timber joists and decking, the dead load is calculated as follows:

$$\begin{aligned} g_{slab} &= (V_{joists} + V_{planks}) \cdot \rho_k \\ &= \left(0.12\ \text{m} \times 0.12\ \text{m} \times 5\ \text{m} \times 21 + 0.022\ \text{m} \times 60\ \text{m}^2\right) \cdot 3.8\ \text{kN}/\text{m}^3 \\ &= 10.7\ \text{kN} \end{aligned}$$

where V_{joists} is the volume of the 21 joists resting on the top plates, being 5 m long, V_{planks} is the volume of the ceiling connected to the joists, being 22 mm thick and ρ_k is the specific weight of the timber, according to Table A1.1.

In some cases, additional earth is inserted between the decking, and this weight is considered as:

$$g_{slab\ infill} = V_{infill} \cdot \rho_c = (12\ \text{m} \times 0.08\ \text{m} \times 0.03\ \text{m} \times 50) \cdot 16\ \text{kN/m}^3 = 23\ \text{kN}$$

where the volume was calculated as 50 rows (10 rows per m) multiplied by the length (12 m), depth (0.08 m) and width (0.03 m) of each infill between the timber elements of the decking, and ρ_c is the specific weight of the timber, according to Table A1.1.

A1.2.1.3 Walls

For the walls, both the timber skeleton and the masonry infill are calculated and taken into account, depending on their specific weights. The walls include timber posts, mudsills, top plates, and braces (all made of the same timber species) and are infilled with mud brick masonry.

Thus:

$$g = Volume \times Specific\,weight = 180\,\text{kN} \tag{Eq. 3}$$

$$g_{timber} = 2.1\,\text{m}^3 \times 3.8\,\text{kN/m} = 8\,\text{kN}$$

$$g_{masonry} = 12.3\,\text{m}^3 \times 14\,\text{kN/m} = 172\,\text{kN}$$

The total deadload resulted as *248 kN*.

A1.2.2 Live Load

Live load is only considered for the slab over the ground floor as 3.5 kN/m^2 (it is rather high, due to the possibility that the owners will store relatively heavy things in the attic area) multiplied by the total area of the house (5×12 m) and results in *210 kN*.

A1.2.3 Snow Load

The Romanian design code [4] was used hereby and the snow load was calculated by the following equation:

$$S = s_k \times C_e \times C_t \times \gamma_{is} \times \mu_1 \tag{Eq. 4}$$

s_k = 2 kN/m² is the characteristic value of the snow load on the roof according to [4].

γ_{is} = 1 is the important factor taking into account the building class (in this case is III) according to CR-1-1-3 (Table 4.1).

C_e = 1 according to CR-1-1-3 (Table 4.2).

C_t = 1 is a thermal coefficient, which takes into account heat loss through the roof; we ignore the losses.

μ_1 = is the shape coefficient, taking into account the shape of the roof (i.e. the angle) and thus:

For $\alpha = 35°$, $\mu_1 = 0.8(60°-35°)/30° = 0.66$ and $\mu_2 = 1.6$

Finally,

$S_{i1} = 1\times1\times1\times2\times0.66 = 1.32$ kN/m²

$S_{i2} = 1\times1\times1\times2\times1.66 = 3.2$ kN/m²

A1.2.4 Wind Load

The Romanian design code [5] was used hereby and the wind load was calculated by the following equation:

$$W_e = q_p(Z_e)\cdot C_{Pe}\cdot C_t\cdot \gamma_{1w}\cdot \mu_1 \qquad \text{(Eq. 5)}$$

$Z_0 = 0.3$ m

$Z_{min} = 5$ m

$$q_p(Z) = C_e(Z)\cdot q_b$$

q_b= 0.6 kPa (kN/m²)

$$C_e(Z) = C_{pq}(Z)\cdot C_r^2(Z)$$

Or

$$C_r^2(Z) = \begin{cases} K_r^2(Z_0)\cdot \ln\dfrac{z}{z_0}\ \ for\, z_{min} \le z \le z_{max} = 200 \\ C_r^2(z = z_{min})\ for\, z \le z_{min} \end{cases}$$

$$K_r^2(Z_0) = 0.189\cdot\left(\frac{Z_0}{0.05}\right)^{0.07} = 0.046\ (for\ cathegory\ III\ of\ soil)$$

$$C_r^2(Z) = 0.046 \cdot \left[ln \frac{5}{0.3} \right]^2 = 0.364$$

$$C_{pq}(Z) = 1 + 7 \cdot I_v(z) \text{ according to [5] (eq. 2.17)}$$

$$I_v(z) = \begin{cases} \dfrac{\sqrt{\beta}}{2.5 \cdot \ln\left(\dfrac{z}{z_0}\right)} & for\, z_{min} \le z \le z_{max} = 200 \\ I_v(z = z_{min})\, for\, z \le z_{min} \end{cases}$$

$$\sqrt{\beta} = 2.35(for\ cathegory\ III\ of\ soil)$$

$$z = 5 \text{ m}$$

$$I_v(z) = \frac{2.35}{2.5 \cdot \ln\left(\dfrac{5}{0.3}\right)} = 0.34$$

$$C_{pq}(Z) = 1 + 7 \cdot 0.34 = 3.38$$

$$C_e(Z) = 3.38 \cdot 0.364 = 1.23$$

$$q_p(Z) = 1.23 \cdot 0.6 = 0.738 \text{ kN / m}^2$$

$$\gamma_{1w} = 1$$

$$q_p(Z) = q_p(Z_e)\ (h \le b_{cladire}$$

ZONE	**F**	*G*	*H*	*I*	*J*
$C_{pe,10}$	**-1.1**	-0.8	-0.8	-0.6	-0.8

C_{Pe} results as the most unfavorable situation, so zone F, value -1.1, this means that only suction will be produced in the area F.

$$W_e = 1 \cdot (-1.1) \cdot 0.738 = 0.362\ kN / m^2$$

A1.2.5 Estimated Natural Period

$$T_1 = C_T h^{3/4} \qquad \text{(Eq. 6)}$$

For timber buildings

$C_T = 0.05$

H = 4 m (total height)

Thus, T1 $= 0.05 \times 4^{3/4} = 0.14$ s

T1 = 0.14 s

A1.2.6 Seismic Load

Considering the position of the house and the seismic zone map given in [6] (Romania's national building code) the following coefficients are adopted:

$a_g = 0.35$ g

$T_B = 0.32$ s

$T_C = 1.6$ s

$T_D = 2$ s

$\gamma_I = 1.0$

The behavior factor ***q*** was considered 2 for timber framed masonry structures [7].

$0 \le T \le T_B$ so

F_b (base shear)=lm $S_{d(T1)}$

$$\beta(T) = 1 + \frac{\beta_0 - 1}{T_B} \cdot T = 1 + \frac{2.5 - 1}{0.32} \cdot 0.14 = 1.66$$

$S_{d(T1)} = a_g \times \beta(T)/q = 0.35 \times 1.66/2 = 0.29$

F_b (base shear) = 248 kN × 0.29 = 72 kN

So the net horizontal force is 29% of the total building weight.

A1.2.7 Load Cases

GP – dead load (self-weight).

PCF – quasi-permanent load (floors, non-structural components, facades) which are variable loads and can be gypsum board walls which can be placed now, but in the future dismantled or the other way around. These loads can lack in the case of the *paianta* house.

U – live load (people, furniture)

SXP, SXN – seismic load on X direction, with and without accidental eccentricity 0.05 Ly..

SYP, SYN – seismic load on Y direction, with and without accidental eccentricity 0.05 Lx

A1.2.8 Load Combinations

The basis of design standard (CR0:2012/Eurocode 0) [8] and timber design standard SR EN 1995 (Eurocode 5) [6], [9], give the following load combination possible:

Ultimate limit state (ULS) without seismic load

$1.35\ P + 1.5\ Sk_1$

$1.35\ P + 1.5\ Sk_2\ (Sk_1) + 1.05\ w(z)$

$1.35\ P + 1.5\ w(z) + 1.05\ Sk_2\ (Sk_1)$

Ultimate limit state (ULS) with seismic load:

$1.35\ P + 0.4\ Sk_1 + \text{Seism}$

Service limit state (SLS):

$1.0\ P + 1.0\ Sk_2\ (Sk_1) + 1.0\ w(z)$

$1.0\ P + 0.4\ Sk_1 + \text{Seism}$

Thus, in SAP2000, the following load combinations were considered:

GF – fundamental group = GP×1.35 + PCF×1.35 + U×1.5 (gravitational, so all the coefficients are maximum).

GS – special group (vertical loads) = GP×1.0 + PCF×1.0 + U×0.4.

GSXP – special group including the seismic load = GS + SXP.

GSXN – special group including the seismic load = GS + SXN.

GSYP – special group including the seismic load = GS + SYP.

GSYN – special group including the seismic load = GS + SYN.

INF – envelope of the combinations GF, GSXP, GSXN, GSYP and GSYN.

A1.2.9 Seismic Design Forces

In SAP2000 the dead load is considered automatically, based on the materials' properties (Young's modulus and cross-section).

For the live load, we considered 3.5 kN/m^2 on the slab over the ground floor, taking into account the deposits that people may have in the attic of their houses. It was applied uniformly distributed on the top plates and considered as belonging to the U load case.

The roof's weight, as it wasn't modeled in SAP2000, was just considered as a point load, applied on each post, thus 34 kN (cover and roof framework) +11 kN (slab over the ground floor) = 45 kN, divided by 18 columns results in *2.55 kN* per each column, the load coming from dead load ($G_{k,\ column}$).

The *snow load is 3.2 kN/m²*, and the *wind load is 0.362 kN/m²*, and they were also uniformly distributed on the top plates.

The *weight of the infills of the walls* are considered supplementary loads, applied also as point loads, only on the columns. Thus 178 kN divided between 18 posts, each post gets a *9.88 kN* point load, placed in the connection with the diagonal.

The modal analysis gave a fundamental period T1 = 0.17 s, which is a little bigger than the calculated one, 0.14 s.

A1.3 DESIGN VERIFICATIONS

Application of the theory to the case study

Most unfavorable loads combination is selected in SAP2000 to make the verifications of the elements meaning: $1.35\ P + 1.5\ Sk2\ (Sk1) + 1.05\ w(z)$

A1.3.1 Verification of Elements (Columns) Loaded in Axial Compression

a. Compression parallel to the fiber

Theory

- When buckling does not occur (**$\lambda_{rel} \leq 0.5$**)

$$\sigma_{c,0,d} \leq f_{c,0,d} \qquad \text{(Eq. 7)}$$

- When buckling occurs

$$\frac{\sigma_{c,0,d}}{k_c \cdot f_{c,0,d}} \leq 1.0 \qquad \text{(Eq. 8)}$$

Where $\sigma_{c,0,d}$ is the normal design compression stress parallel to the fiber and is given by:

$$\sigma_{c,0,d} = \left(\gamma_G \cdot F_G + \gamma_Q \cdot F_Q\right) / A_n \qquad \text{(Eq. 9)}$$

$f_{c,0,d}$ is the design strength of timber in compression parallel to the fiber and is calculated based Eq. 1

F_G, F_Q = axial forces resulting from the permanent loads (G), and variable (Q);

γ_G, γ_Q – safety coefficients corresponding to loading conditions, taken as 1.25 herein;

A_n – net area of the element;

k_c – coefficient considering the buckling and calculated as follows:

$$k_c = \frac{1}{\left(k + \sqrt{k^2 - \lambda_{rel}^2}\right)} \qquad \text{(Eq. 10)}$$

where

$$k = 0.5 \cdot \left[1 + \beta_c\left(\lambda_{rel} - 0.5\right) + \lambda_{rel}^2\right]$$

β_c – coefficient taking into account the element's imperfections and has a value of 0.2 for massive timber.

λ_{rel} – relative slenderness calculated with:

$$\lambda_{rel} = \sqrt{\frac{f_{c,0,k}}{\sigma_{c,crt}}} \qquad \text{(Eq. 11)}$$

The critical effort, $\sigma_{c,crt}$ is calculated with:

$$\sigma_{c,crt} = \frac{\pi^2 \cdot E_{0.05}}{\lambda^2} \qquad \text{(Eq. 12)}$$

Application to the case study

Based on material characteristics (compression strength in the direction of the fibers = 40 MPa) we have:

$$f_{c,0,d} = \frac{1.1 \times 40 \text{ MPa}}{1.3} = 33.85 \text{ N/mm}^2$$

From the SAP2000 model, for the ULS combination without earthquake we have:
The maximum axial load obtained in a column is N = 52 kN

Thus, $\sigma_{c,0,d} = \frac{\gamma_c \cdot N}{A_n} = \frac{1.25 \cdot 52}{120 \cdot 120} = 3.7\,\mathrm{N/mm^2}$

It results that: $\sigma_{c,0,d} \leq f_{c,0,d}$ *OK!*
Then, the buckling verification is done:

$$\sigma_{c,crt} = \frac{\pi^2 \cdot E_{0.05}}{\lambda^2} = \frac{3.14^2 \times 8900}{64^2} = 21.4\,\mathrm{N/mm^2}$$

$$\lambda = \frac{l_f}{i} = \frac{l_f(3)}{\sqrt{\frac{I}{A}}} = \frac{1 \times 2.17\mathrm{m}}{17280000\ \mathrm{mm^4}} = 64 < \lambda_{adm} = 120\,OK!$$

$$\lambda_{rel} = \sqrt{\frac{f_{c,0,k}}{\sigma_{c,crt}}} = \sqrt{\frac{40\,\mathrm{N/mm^2}}{21.4\,\mathrm{N/mm^2}}} = 1.36$$

$$k = 0.5 \cdot \left(1 + 0.2(1.36 - 0.5) + 1.36^2\right) = 1.51$$

$$k_c = \frac{1}{\left(k + \sqrt{k^2 - \lambda_{rel}^2}\right)} = \frac{1}{\left(1.51 + \sqrt{1.51^2 - 1.36^2}\right)} = 0.461$$

$$\frac{\sigma_{c,0,d}}{k_c \cdot f_{c,0,d}} = \frac{3.6\,\mathrm{N/mm^2}}{0.461 \times 33.85\,\mathrm{N/mm^2}} = 0.23 < 1.0\,OK!$$

b. Compression oblique to the fiber

Theory

$$\sigma_{c,\alpha,d} \leq \frac{f_{c,0,d}}{\left(\frac{f_{c,0,d}}{f_{c,90,d}} \cdot \sin^2\alpha + \cos^2\alpha\right)} \qquad \text{(Eq. 13)}$$

where $\sigma_{c,\alpha,d}$ is the normal design stress of timber in compression oblique to the fiber;

$f_{c,0,d}$ *and* $f_{c,90,d}$ are the design strengths of timber in compression parallel and perpendicular to the fiber, respectively and are calculated based on Eq. 1.

Application to the case study

Based on material characteristics (compression strength perpendicular to the fiber direction – 5.8 MPa) $f_{c,0,d} = 4.9$ MPa.

Then, from the SAP2000 model, the ULS combination with earthquake (ULS +SYN), the maximum shear force on the column, coming from the diagonal (at 42°), was obtained as 13.5 kN.

$$\sigma_{c,\alpha,d} = \frac{1.25 \cdot 13.5\,\text{kN}\ (\textit{from SAP2000})}{120 \times 120} = 0.96\,\text{N}/\text{mm}^2$$

$$\frac{f_{c,0,d}}{\left(\frac{f_{c,0,d}}{f_{c,90,d}} \cdot \sin^2\alpha + \cos^2\alpha\right)} = \frac{33.85}{\frac{33.85}{4.9 \times} \cdot \sin^2 42° + \cos^2 42°}$$
$$= 4.03\,\text{N}/\text{mm}^2 > 0.96\,\text{N}/\text{mm}^2 OK!$$

A1.3.2 Verification of Elements (Columns) Loaded in Shear

Theory

$$\tau_d = (\gamma_G \cdot T_G + \gamma_Q \cdot T_Q) \cdot S_x / b \cdot I_x \leq f_{v,d} \qquad \text{(Eq. 14)}$$

where τ_d is the tangential design stress;

$f_{v,d}$ is the design strength of timber in shear and is calculated based the Eq. 1;

T_G, T_Q = shear forces resulting from the permanent loads (G), and variable (Q);

γ_G, γ_Q – safety coefficients corresponding to loading conditions, taken as 1.25 herein;

S_x, I_x – static moment and inertia moment, respectively, of the cross-section with respect to the neutral axis;

b – the thickness of the cross-section.

Application to the case study

The maximum shear force is obtained from the SAP2000 model, in the combination ULS without earthquake, and it resulted as 24.7 kN in one of the intermediary columns.

Thus,

$$\tau_d = \frac{\gamma \cdot T_{SAP}}{A_n} = \frac{1.25 \cdot 25\,\text{kN}}{120 \times 120} = 2.17\,\text{N/mm}^2$$

Based in material characteristics (shear strength of the timber is 6.7 N / mm^2):

$$f_{v,d} = \frac{1.1 \times f_{v,k}}{1.3} = 5.66\,\text{N/mm}^2$$

2.17 N / mm^2 < 5.66 N / mm^2 *OK!*

A1.3.3 Verification of Elements (Columns) Loaded in Compression and in Bending

Theory

For elements with a slenderness $\lambda_{rel} \leq 0.5$ the following equations must be satisfied:

$$\left(\frac{\sigma_{c,0,d}}{f_{c,0,d}}\right)^2 + \frac{\sigma_{m,x,d}}{f_{m,x,d}} + k_m \cdot \frac{\sigma_{m,y,d}}{f_{m,y,d}} \leq 1 \qquad \text{(Eq. 15)}$$

$$\left(\frac{\sigma_{c,0,d}}{f_{c,0,d}}\right)^2 + k_m \cdot \frac{\sigma_{m,x,d}}{f_{m,x,d}} + \frac{\sigma_{m,y,d}}{f_{m,y,d}} \leq 1 \qquad \text{(Eq. 16)}$$

where $\sigma_{c,0,d}$ is the unit compression stress calculated with Eq. 9;

$\sigma_{m,x,d}$ *and* $\sigma_{m,y,d}$ are the design unit bending stress for axis x and y;

$f_{c,0,d}$ is the design strength of timber in compression parallel to the fiber and is calculated based on Eq. 1;

$f_{m,x,d}$ *and* $f_{m,y,d}$ are the design strength of timber in bending parallel to the fiber and is calculated based on Eq. 1;

k_m – coefficient considering the shape of the cross-section, and for rectangular ones it is 0.7.

For elements whose slenderness $\lambda_{rel} > 0.5$ the following equations apply:

$$\frac{\sigma_{c,0,d}}{k_{c,x} \cdot f_{c,0,d}} + \frac{\sigma_{m,x,d}}{f_{m,x,d}} + k_m \cdot \frac{\sigma_{m,y,d}}{f_{m,y,d}} \leq 1 \quad \text{(Eq. 17)}$$

$$\frac{\sigma_{c,0,d}}{k_{c,y} \cdot f_{c,0,d}} + k_m \cdot \frac{\sigma_{m,x,d}}{f_{m,x,d}} + \frac{\sigma_{m,y,d}}{f_{m,y,d}} \leq 1 \quad \text{(Eq. 18)}$$

where $k_{c,x}$ *and* $k_{c,y}$ are the coefficients taking into account the buckling with respect to axis x and y, calculated with $k_c = \frac{1}{\left(k + \sqrt{k^2 - \lambda_{rel}^2}\right)}$ (Eq. 10)

Application to the case study

As calculated above, $\lambda_{rel} = 1.36$, thus we are in the second situation, and the values for the terms in Eq. 18 and 19 are as follows (calculated above):

$$\sigma_{c,0,d} = 3.6 \mathrm{N} / \mathrm{mm}^2$$

$$f_{c,0,d} = 33.85 \mathrm{~N} / \mathrm{mm}^2$$

$$k_{c,x} = k_{c,y} = 1.51$$

$$f_{m,x,d} = f_{m,x,d} = \frac{1.1 \times 61 \left(\textit{bending strength of timber}\right)}{1} = 67.1 \mathrm{N} / \mathrm{mm}^2$$

The maximum bending moment's corresponding force is taken from the SAP2000 model, within the load combinations ULS without earthquake, and obtained 25 kN.

$$\sigma_{m,x,d} = \sigma_{m,y,d} = \frac{\textit{Maximum bending moment's corresponding force in SAP2000}}{120 \times 120} = \frac{25000 \mathrm{N}}{14400 \mathrm{mm}^2} = 1.74 \mathrm{N} / \mathrm{mm}^2$$

Thus:

$$\frac{3.6}{1.51 \cdot 33.85} + \frac{1.74}{67.1} + 0.7 \cdot \frac{1.74}{67.1} \leq 1 OK!$$

A1.3.4 Verification of Elements (Beams) Loaded in Bending

a. Strength limit state verification

Theory

$$\sigma_{m,d} \leq f_{m,d} \qquad \text{(Eq. 19)}$$

where $\sigma_{m,d}$ is the stress given by the design moment;
$f_{m,d}$ – is the design strength of the timber in bending determined from Eq. 1

But in some situations, such as this one, the Eq 1 can be modified and extra coefficients can be added, taking into account more influencing factors:

$$f_{m,d} = k_{mod} \cdot k_{crit} \cdot k_{1s} \cdot k_h \cdot f_{m,k} / \gamma_M \qquad \text{(Eq. 20)}$$

k_{crit} – is the coefficient taking into account the instability of the element;

$$k_{crit} = \begin{cases} 1 \ when \ \lambda_{rel,m} \leq 0.75 \\ 1.56 - 0.75\lambda_{rel,m} \quad when 0.75 \leq \lambda_{rel,m} \leq 1.4 \\ \dfrac{1}{\lambda^2_{rel,m}} \quad when \ \lambda_{rel,m} \geq 1.4 \end{cases} \qquad \text{(Eq. 21)}$$

$$\lambda_{rel,m} = \sqrt{\frac{f_{m,k}}{\sigma_{m,crt}}} \qquad \text{(Eq. 22)}$$

where $\sigma_{m,crt}$ is the critical stress determined for $E = E_{0.05}$ and taking into account the factor *m*, to equivalate the bending moment, as in Table 4.17 [2]. In this case, it was considered 0.57, as the static scheme was considered fixed at one end and pinned at the other.

$$\sigma_{m,crt} = \frac{0.75 \cdot E \cdot b^2}{h \cdot l_{ef}} \quad \text{(Eq. 23)}$$

k_{1s} is the coefficient taking into account the assembly's effect over the capacity of the element, and is taken as 1;
k_h is the coefficient taking into account the height and is calculated with:

$$k_h = min\begin{cases} \left(\frac{150}{h}\right)^{0.2} \\ 1.3 \end{cases}$$

Application to the case study

The main beam element is the top plate, bearing the roof loads and resting on the posts and masonry infill. The mudsill is continuously supported on the foundations (or socle), and thus is not subjected to bending. For simplicity, the infill is not considered herein, and the top plate is calculated as resting only on the posts.

$$\sigma_{m,crt} = \frac{0.75 \cdot E \cdot b^2}{h \cdot l_{ef}} = \frac{0.75 \times 8900 \times 120^2}{120 \times 2170} = 369.12 \text{ N / mm}^2$$

$$\lambda_{rel,m} = \sqrt{\frac{f_{m,k}}{\sigma_{m,crt}}} = \sqrt{\frac{61 \text{ MPa}}{369.12}} = 0.4 \Rightarrow k_{crit} = 1$$

$$k_h = min\begin{cases} \left(\frac{150}{h}\right)^{0.2} = 1.04 \\ 1.3 \end{cases}$$

$$f_{m,d} = k_{mod} \cdot k_{crit} \cdot k_{1s} \cdot k_h \frac{f_{m,k}}{\gamma_M} = 1.1 \times 1 \times 1 \times 1.04 \times \frac{61}{1.3} = 53.97 \text{ N / mm}^2$$

The maximum moment in the top plates is considered from the SLS without earthquake load combination (on the transversal top plates), and from the SAP2000 model was obtained as 3.73 kNm:

$$\sigma_{m,d} = \frac{3.73\times10^6}{W} = \frac{3730000\ \text{Nmm}}{17280000\ \text{mm}^4} = 12.8\ \text{N}/\text{mm}^2$$

where W is the section modulus, and for this 120 x 120 mm cross-section is 288000 mm^3.

$$\sigma_{m,d} \leq f_{m,d}\ OK!!$$

b. Deformation limit state verification

Theory

The final deformation under service limit state verification is calculated using the following formula:

$$u_{fin} = u_{inst}\left(1+k_{def}\right) \quad \text{(Eq. 24)}$$

where u_{inst} is the instantaneous deformation under loading coming from the most unfavorable combination;

k_{def} – is the coefficient taking into account the deformation in time under the effect of creep and moisture (Table 4.22 in [2]), hereby taken as 0.6.

$$u_{inst} = \frac{5}{384}\cdot\frac{q^{SLS}\cdot L^4}{E\cdot I} \quad \text{(Eq. 25)}$$

Application to the case study

The load from the SLS (wind, live and snow) was obtained as 7 kN/m and in a first step we consider the opening of the plate, between the supports as 3.36 m (see Figure A1.1). Thus:

$$u_{g,inst} = \frac{5}{384}\frac{7\times3360^4}{8900\times17280000} = 74\ \text{mm}$$

$$u_{g,fin} = u_{g,inst}\left(1+k_{def}\right) = 74\times1.6 = 120\,mm \geq u_{adm,fin} = \frac{l}{200} = \frac{2170}{200} = 10.85\,\text{mm}$$

NOT OK!!

But in reality, the transversal top plate is also supported by vertical timber elements placed on the edges of the windows (see Figure A1.1). And the distance between the supports, within the static scheme would be maximum 1.28 m, and if taken as the opening between the supports, the deflection would be 1.59 mm, and then $u_{g,fin}$ would meet the deflection requirements. Moreover, in reality, the masonry infill also carries the loads, and thus the beam's deflection is additionally reduced.

A1.3.5 Verification of Elements (Beams) Loaded in Shear

Theory

$$\tau_d = \left(\gamma_G \cdot T_G + \gamma_Q \cdot T_Q\right) \cdot S_x / b \cdot I_x \leq f_{v,d} \quad \text{(Eq. 26)}$$

where τ_d is the tangential design stress;

$f_{v,d}$ is the design strength of timber in shear and is calculated based on Eq. 1;

T_G, T_Q =shear forces resulting from the permanent loads (G), and variable (Q);

γ_G, γ_Q – safety coefficients corresponding to loading conditions, and is taken hereby as 1.25;

S_x, I_x – static moment and inertia moment, respectively, of the cross-section with respect to the neutral axis;

b – the thickness of the cross-section.

Application to the case study

The shear force was considered from the ULS without earthquake and was obtained from the SAP2000 model as 7.4 kN.

$$\tau_d = \frac{\gamma \cdot T_{SAP}}{A} = \frac{1.25 \cdot 7.4\,\text{kN}}{120 \times 120} = 0.64\,\text{N}/\text{mm}^2$$

$$f_{v,d} = \frac{1.1 \times f_{v,k}}{1.3} = 5.66\,\text{N}/\text{mm}^2$$

$0.64\ \text{N}/\text{mm}^2 < 5.66\ \text{MPa}$ *OK!*

A1.3.6 Verification of Elements (Beams) Loaded in Compression Perpendicular to the Fiber Direction

Theory

$$\sigma_{c,0,d} \leq k_{c,90} \cdot f_{c,90,d} \qquad \text{(Eq. 27)}$$

where $k_{c,90}$ is a coefficient taking into account the way the compression is applied and is given in Table 4.16 [2];

$f_{c,90,d}$ – is the compression strength perpendicular to the grain determined with Eq. 1.

Application to the case study

$$f_{c,90,d} = \frac{1.1 \times 5.8\,MPa}{1.3} = 4.9\,\text{N}/\text{mm}^2$$

$$k_{c,90} = \frac{l + (150 - l)}{170} = \frac{120 + 30}{170} = 0.88$$

$$\sigma_{c,0,d} = 3.7\text{N}/\text{mm}^2$$

$$0.88 \times 4.9 = 4.3\,\text{MPa} > 3.7\,\text{N}/\text{mm}^2 OK!$$

A1.3.7 Verification of Joints

A. Compression perpendicular to the fiber direction

Theory

$$\sigma_{c,90,d} = \frac{c}{b \cdot l} \leq k_{c,90} \cdot f_{c,90,d} \qquad \text{(Eq. 28)}$$

where $\sigma_{c,90,d}$ – compression stress perpendicular to the fiber;

$k_{c,90}$ is a coefficient taking into account the way the compression is applied and is given in Table 4.16 [2];

$f_{c,90,d}$ – is the compression strength perpendicular to the grain determined with Eq. 1

V_d – shear force;

b×l – compressed area.

Application to the case study

The shear force is obtained from the SAP2000 model, joints diagram, load combination SLU without earthquake, and is maximum 57 kN.

$$\sigma_{c,90,d} = \frac{V_d}{b \times l} = \frac{57(shear\ force\ from\ SAP2000)}{12800\,\text{mm}^2} =$$

$$= 4.45\,\text{N}/\text{mm}^2 \leq 0.88 \times 4.9\,\text{N}/\text{mm}^2 = 4.3\,\text{N}/\text{mm}^2\,NOT\ OK!!$$

Thus, the section of the columns should be increased.

However, the model assumes that the masonry infill's weight is concentrated in the posts, in the connections with the diagonal braces. This assumption was made in order to take into account the weight of the masonry infill, within the modal analysis. So, in reality, the post doesn't carry so much load perpendicular to the fiber's direction on the mudsill. Anyway, for the sake of the correct calculation, in this step, the posts' cross-section should be increased, at least for the maximum loaded joints (axis intersections 4A and 4D from Figure A1.1) to a 140 x 140 mm cross-section with a minimum tenon of 78 x 78 mm.

b. In shear perpendicular to the fiber direction

Theory

To verify the shear perpendicular to the fiber direction the following equation is used:

$$\tau_{d\,max} = 1.5 \cdot V_d / (b \cdot h_e) \leq k_v f_{v,d} \qquad \text{(Eq. 29)}$$

where b×h_e is the shear area;

k_v – is a coefficient depending on the connection geometry and depends on the height of the element (h), height of the mortise cross-section (h_e) and distance X.

For the central mortise the following equation is used:

$$k_v = min\begin{cases} 1 \\ \dfrac{5}{\left[\sqrt{h}\cdot\left(\sqrt{\dfrac{\dfrac{h_e}{1-\dfrac{h_e}{h}}}{h}}\right)+0.8\cdot\dfrac{\sqrt{\dfrac{h}{h_e}-\left(\dfrac{h_e}{h}\right)^2}}{h}\right]} \end{cases} \quad \text{(Eq. 30)}$$

Application to the case study

$h = 120$ mm

$h_e = 80$ mm

$$\left[\sqrt{120}\cdot\left(\sqrt{\frac{\frac{80}{1-\frac{80}{120}}}{120}}\right)+0.8\cdot\frac{\sqrt{\frac{120}{80}-\left(\frac{80}{120}\right)^2}}{120}\right]$$

$$=\left[\sqrt{120}\cdot\left(\sqrt{\frac{\frac{80}{1-\frac{80}{120}}}{120}}\right)+0.8\cdot\frac{\sqrt{\frac{120}{80}-\left(\frac{80}{120}\right)^2}}{120}\right]$$

$$=10.9544(1.42)+0.0068=15.33$$

Thus:

$k_v = 1$

$$f_{v,d} = 5.66\,\mathrm{N/mm^2}$$

$$b \cdot h_e = 120 \times 80 = 9600\ \mathrm{mm^2}$$

$$\tau_\mathrm{d} = 1.5 \times \frac{57000\,\mathrm{N}}{120 \times 80} = 8.9 \frac{\mathrm{N}}{\mathrm{mm^2}} \leq 5.66 \frac{\mathrm{N}}{\mathrm{mm^2}} \quad \textit{NOT OK!!}$$

The same situation as in the previous step, and if the cross-section is increased to 140x140 for the post and 110 x 110 the tenon, the result will be τ_d = $5.55\,\mathrm{N/mm^2}$, and verifies.

The diagonal braces' connection to the post is in this case done as a simple nailed connection, thus the only resistance is in the two nails, but in reality, it is neglectable. For very low forces, the nails will bend, and the brace will slide up and down on the post, depending on the direction of the loading.

REFERENCES

[1] A. Dutu, M. Niste, I. Spatarelu, and D. I. Dima, "TFMRO project (Seismic evaluation of Romanian traditional residential buildings)." [Online]. Available: http://tfmro.utcb.ro/

[2] C. Furdui, *Constructii din lemn: materiale si elemente de calcul.* Timisoara: Politehnica, 2005.

[3] Ministry of Regional Development and Public Administration (MDRAP), Technical *S*pecification *R*egarding the *P*rotection of *T*imber *S*tructures' *E*lements against *A*gressive *F*actors. *R*equirements, *P*erformance *C*riteria and *P*revention and *M*itigation *M*easures – ST 049-2014. MO 318, 2015.

[4] Ministry of Regional Development and Public Administration (MDRAP), Design *C*ode. Evaluation of *S*now *L*oad for *C*onstructions. 2012.

[5] Ministry of Regional Development and Public Administration (MDRAP), Design *C*ode. Evaluation of *W*ind *L*oad for *C*onstructions. 2012.

[6] Ministry of Regional Development and Public Administration (MDRAP), *P100-1/2013. Cod de proiectare seismică – Partea I – Prevederi de proiectare pentru clădiri [Seismic Design Code – Part I: Design Prescriptions for Buildings].* Monitorul Oficial al României, No. 558 bis/3 September 2013 (in Romanian), 2019.

[7] A. Ceccotti, M. Massari, and L. Pozza, "Procedures for seismic characterization of traditional and modern wooden building types," *International Journal for Quality Research*, vol. 10, no. 1, pp. 47–70, 2016.

[8] Ministry of Regional Development and Public Administration (MDRAP), Design *C*ode. Construction *D*esign *B*asis. 2012.
[9] CEN (European Committee for Standardization), Eurocode 5 – Design of *T*imber *S*tructures – Part 1-1: General – Common rules and *R*ules for *B*uildings. 2004.

Annex 2
Design Example of a Traditional Japanese Residential House

A2.1 BUILDING OUTLINE

A house model was selected, as simple as possible, from the Calculation Guideline for Traditional Wooden Building Structures [1] just to show and understand the principles of the Japanese way to design new traditional timber houses.

The location of the house is corresponding to the Class 1 soil type and surface roughness classification IV, according to the Japanese building code.

The total floor area of the house is 79.50 m^2, building area: 221.17 m^2, having two stories, a total height of 7.9 m and eave height: 5.98 m. The house's layout and plan are shown in Figure A2.1.

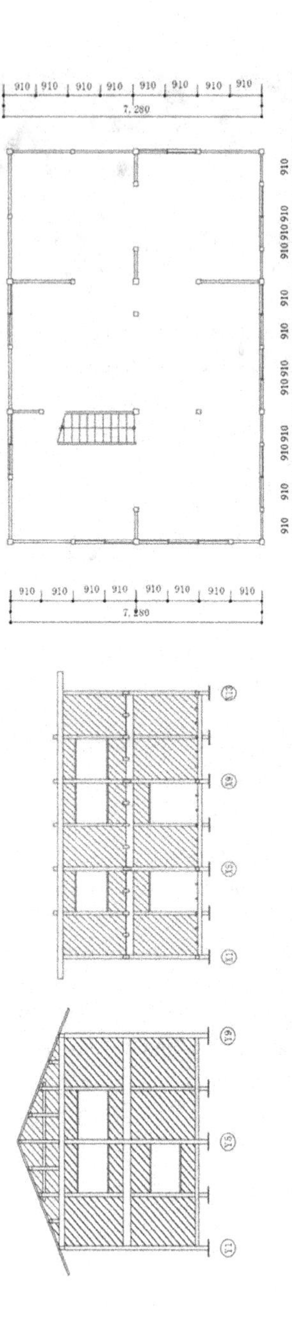

FIGURE A2.1 Building layout and floor plan.

A2.2 STRUCTURAL CHARACTERISTICS

The foundations under the posts are built of stone, but horizontal and vertical movement is restrained by anchor bolts. The vertical structural components that resist horizontal forces are the full-length mud walls, small mud walls, posts with an overhang, foundations, and timber connections. The wall thickness of the mud wall is 65 mm. The main joints are the mortise-and-tenon type. One thing to be mentioned it the standard module of Japanese walls, which has a length of 910 mm, and each wall is made of a certain number of modules.

A2.3 STRUCTURAL DESIGN (CALCULATION) PRINCIPLES

The Japanese building codes have three concepts: (a) the permanent and imposed loads to be safely supported, without excessive deformations; (b) in the case of a rare medium-scale hazard (earthquake, snowfall, windstorm, etc.) the building should not exhibit any damage; (c) in the case of an extremely rare large-scale hazard, the building should not collapse.

First, the category the building belongs to is identified, and the small buildings are found in category IV: wooden buildings, with the number of stories lower or equal to two, a total floor area lower or equal to 500 m^2, a building height lower or equal to 13 m and an eave height lower or equal to 9 m.

The design method corresponds to the category the building belongs to, and Figure A2.2 shows the design methods recommended by the Building Center of Japan.

Thus, for the category IV buildings, the requirements are only to comply with the structural specifications. Structural calculations are not required. But one can use structural combinations, and for *minka* houses with a greater importance, the response and limit capacity calculation can be done. Structural specifications are provided for wooden houses, but they also exist for traditional wooden houses and is called "Guidelines for Structural Calculation of Traditional Wooden Buildings Using Limit Strength Calculations".

The main indications received within these structural specifications are related to the structure of mudsills and foundations, size of posts, necessary strength and quantity of braces as well as structural frames, connection methods, preservative measures, etc.

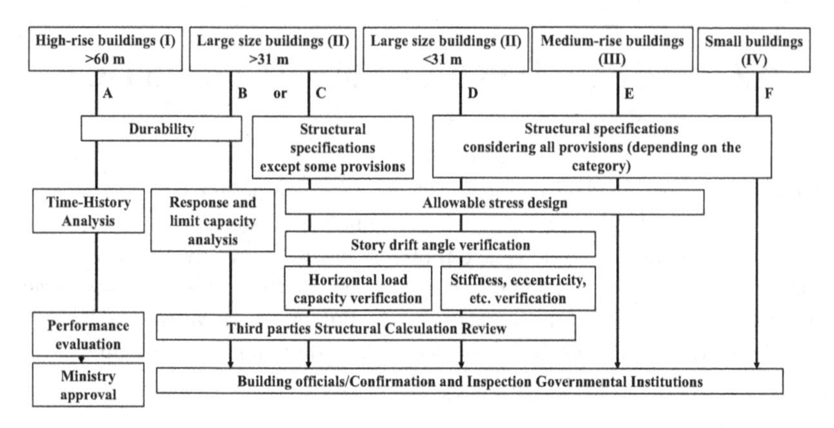

FIGURE A2.2 Design method recommended depending on the category of building in Japan.

The load-deformation relation of the vertical structure is calculated based on the load-deformation relation of the vertical component elements (the horizontal structure is assumed to be a rigid floor).

The design principle is to ensure that the deformation angle does not exceed the safety limit for extremely rare earthquake motion, that the stress of the member is less than or equal to the material strength, and that the deformation of the joint is less than or equal to the ultimate deformation (Chapter 5 in [2]).

The above verification is applied to the joints of structures with load-bearing walls (there are joints only between the posts and beams of this building). Other joints are excluded from the scope of the verification.

The building is of general residential scale, so the description of allowable stress calculations for beams and columns for long-term loads is omitted.

The stresses at the joints at the maximum load carrying capacity of the structure shall not exceed the bearing capacity of the joints.

The *damage limit* deformation angle shall be *1/120 rad* and the *safety limit* deformation angle shall be *1/15* rad or less.

The eccentricity length (e_x and e_y) shall be calculated starting from the center of gravity and center of stiffness. Then, after calculating the rotational stiffness (K_R) around the center of stiffness, we can obtain the radius of elasticity (r_e) and with it, the eccentricity ratio (R_{eX} and R_{eY}), and thus the *Fe*, which is a coefficient depending on the resulted eccentricity.

A2.4 MATERIALS' PROPERTIES

The material is cedar timber (cedar E70), and its physical properties are as follows:

- Compression strength, Fc, 18.0 N/mm^2,
- Tension strength 13.2 N/mm$^{2;}$
- Bending strength 22.2 N/mm^2;
- Allowable stress is 1.1 $F_c/3$ for long-term and 2 $F_c/3$ for short-term.

A2.5 DESIGN LOADS

The permanent and variable loads were considered based on the national values determined and given in standards, similar to the way they are provided for the Romanian *paianta*. Thus, the dead load, live load, wind load, snow load and seismic load are identified in this step.

A2.6 CREATION OF LOAD-DISPLACEMENT RELATIONSHIPS FOR EACH VERTICAL STRUCTURAL ELEMENT

The eccentricity of this building is 0.28 in the EW direction and 0.31 in the NS direction for the first floor, and 0.21 in the EW direction and 0.18 in the NS direction for the second floor. Although for this case the eccentricity slightly exceeded 0.3, the calculation was performed on a mass point system using an eccentricity correction. The horizontal plane element (slab) was assumed as rigid, and the force-displacement curve results from the sum of the force-displacement curves of the vertical elements contributing to the lateral resistance. The torsion is taken into account if it occurs.

Then, it is necessary to express the load-deformation relationship of these vertical structural elements by calculating the load-deformation relationship and applying forces on the vertical structural elements, and then expressing them in a multi-linear format that connects the force values at specific deformations (Figure A2.3).

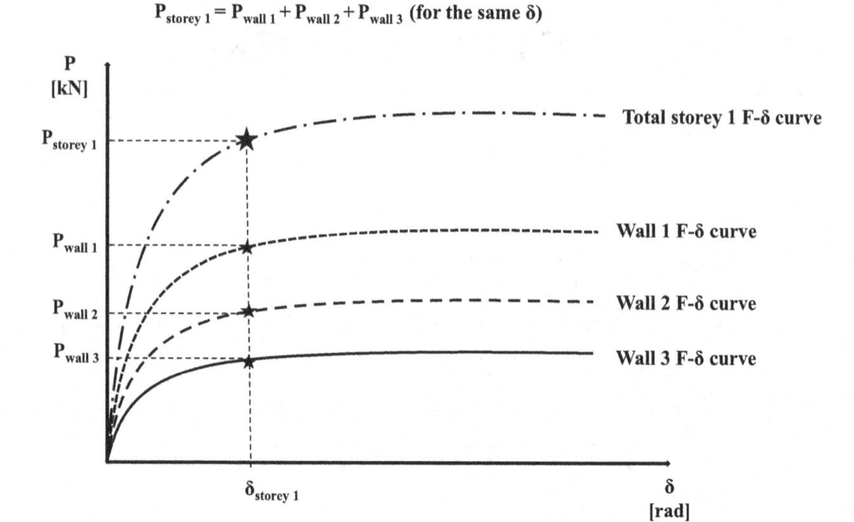

FIGURE A2.3 Method to sum the F-δ curves of the structural walls in a building on X or Y direction.

As a horizontal resistance mechanism, the bearing capacity was added for each specified deformation angle for full walls, free-standing posts with small walls, small walls sandwiched between full walls, and mortise-tenon joints with dowel. For the independent posts with small walls, a load-bearing capacity-deformation curve is created for each post, and the load-bearing capacity at the specified deformation angle is calculated by linear completion. For the 120 mm square columns with small walls (l_w =1820 mm, h_w = 910 mm) on the first floor, bending failure occurs at an inter-story displacement of 84 mm. This value is obtained based on the "Calculation Guideline for Traditional Wooden Building Structures" in Japan, which calculates the total drift composed of the shear deformation of the small wall and the bending deformation of the column. This total drift is also given in the guideline as a function of the lateral force, and thus the force-deformation curve is obtained.

A2.7 LOAD-DISPLACEMENT RELATIONSHIP FOR EACH FLOOR

The skeleton curves of each floor were summed together to create the skeleton curves of the layered (summed) force-displacement curve. Eccentricity is considered as a modified skeleton curve by dividing the skeleton curve of each floor by F_e. In addition, full elasto-plastic replacement is done for each layer to calculate equivalent damping constants, and also yield and limit deformations.

A2.8 RATE OF REDUCTION OF ACCELERATION DUE TO DAMPING

To set the damping constant (F_h) of the building, the load-deformation curve of each floor is replaced by an equivalent fully elastoplastic system, and the equivalent damping constant (h_{ei}) corresponding to the deformation situation of each floor is calculated.

A2.9 CALCULATION OF EARTHQUAKE RESPONSE

The seismic load was calculated in the previous steps, and the response of the building is calculated in terms of displacement increments for rare and extremely rare earthquake levels. The guideline provides a table (Table 10.3-15) where displacements values are given depending on the type of wall considered for the house. Then, using incremental analysis and the Newmark method, the spectrum can be obtained, and check the interaction point between the building's curve on X or Y direction and the spectrum.

A2.10 VERIFICATION FOR EARTHQUAKE LOADS

A2.10.1 Tests at Damage Limit and Safety Limit

A2.10.1.1 Force applied in EW direction

The representative displacement at the damage limit is obtained from analysis using calculation software, and for the example building is 0.0253 m (0.02 m

(1/144 rad) for the first floor and 0.0086 m (1/297 rad) for the second floor). These values are obtained from the intersection of the curves, the building layered one with the response spectra for rare earthquake and the one for extremely rare earthquakes. Therefore, the damage limit deformation angle is not exceeded at the damage limit and is less than or equal to the damage limit deformation angle of the partitions, etc.

The calculation could not be performed at the safety limit, because the curves do not intersect, and this means that the structure will collapse.

A2.10.1.2 Force applied in NS direction

The representative displacement at the damage limit was 0.0251 m (0.0219 m (1/132 rad) for the first floor and 0.0055 m (1/464 rad) for the second floor). Therefore, the damage limit deformation angle is not exceeded at the damage limit and is less than or equal to the damage limit deformation angle of the partitions, etc.

The calculation could not be performed at the safety limit, in this direction either.

The connections are also checked. For rare earthquakes, a verification is made to check if the allowable bearing capacity (allowable stress) is less than or equal to the maximum bearing capacity (material strength) of the connection. For very rare earthquakes, verification is made if the maximum bearing capacity (material strength) of the connection is less than or equal to the maximum bearing capacity (material strength) of the connection.

Tensile forces acting on the main joints should not exceed the maximum horizontal bearing capacity of the adjacent seismic elements. In the present building, the maximum is the 2-ply clay-coated wall (maximum bearing capacity: 12.5 kN). The allowable bearing capacity of the mortise-and-tenon is less than that, so it is safe. These values are provided in the manual and are based on an extensive experimental program.

A2.11 VERIFICATION FOR WIND LOAD

A2.11.1 EW Direction

The wind load is 32.4 kN, which occurs rarely. We then check the force-displacement curve for the first floor, obtained by summation of all the contributing structural elements. In this curve, we check where is the 32.4 kN value, and obtain the corresponding displacement. In this case, the deformation produced by the wind load on the first floor is 0.0124 m (1/232 rad).

The wind load is 51.9 kN, which occurs very rarely. At this time, the deformation produced by the wind load on the first floor is 0.0238 m (1/121 rad).

A2.11.2 NS Direction

The wind load is 47.2 kN, which occurs rarely. At this time, the deformation produced by wind load on the first floor is 0.021 m (1/137 rad).

The wind load is 75.5 kN, which occurs very rarely. At this time, the building collapses. And thus, the calculations should be made again, with a stronger building.

Design examples of other types of timber houses in Japan are also available, and one of the is thoroughly presented in [3]. These design examples are meant to get a better understanding of the Japanese design method, for wood houses, traditional or not.

REFERENCES

[1] Committee of design manual for traditional wooden buildings, *A Manual of Aseismic Design Method for Traditional Wooden Buildings Including Specific Techniques for Unfixing Column Bases to Foundation Stones.* Kyoto, Japan: Gakugei Shuppansha, 2019.

[2] Architectural Institute of Japan, *Recommendation for Structural Calculation of Traditional Wood Buildings by Calculation of Response and Limit Strength.* Kyoto, Japan: Maruzen Publishing Co. Ltd., 2013.

[3] R. Minami *et al.*, "performance verification and trial design for high-rise timber frame buildings with buckling-restrained brace part 2: Analysis of the trial design building," Proceedings of the World Conference on Timber Engineering, pp. 2968–2977, 2023.

Index